Operations Research and Systems Engineering

This book presents an overview of operations research and systems engineering and takes a look into both fields on content, histories, contributions, and future directions so a sound career choice can be made for those who might be deciding on a career path. The book also offers how these two fields can be integrated and used in current times and into the future.

Operations Research and Systems Engineering: Growth and Transformation traces the history of both fields of research as well as offers comments on the importance of both areas of study. By taking a look back with a historical perspective and also looking forward with the presentation of applications currently being used, someone looking to make a sound career choice will be able to decide which area they want to move towards. The book also offers how to integrate both operations research methods with systems engineering concepts and tools and provides a comparison between the two, along with how they can work together in the future.

The goal of this book is to provide the reader with enough information so they can move forward with their career goals. It is also an ideal book that provides engineers, scientists, and mathematicians with a way to broaden their knowledge and areas of study.

T0353595

Operations Research and Systems Engineering
Growth and Transformation
Commentaries on the Profession

Howard Eisner

CRC Press
Taylor & Francis Group
Boca Raton London New York

CRC Press is an imprint of the
Taylor & Francis Group, an **informa** business

First edition published 2023
by CRC Press
6000 Broken Sound Parkway NW, Suite 300, Boca Raton, FL 33487-2742

and by CRC Press
4 Park Square, Milton Park, Abingdon, Oxon, OX14 4RN

CRC Press is an imprint of Taylor & Francis Group, LLC

© 2023 Howard Eisner

ISBN: 978-1-032-30749-7 (hbk)
ISBN: 978-1-032-30781-7 (pbk)
ISBN: 978-1-003-30661-0 (ebk)

DOI: 10.1201/9781003306610

Typeset in Times
by MPS Limited, Dehradun

This book is dedicated to my grandchildren: Rebecca, Gabriel, Jacob

Benjamin and Zachary and my wife, June Linowitz

Contents

Contents

Preface

This book provides an overview of both operations research and systems engineering. As such it contains guidance for the student who is trying to decide which of these two fields to declare as a major in an academic program. Or, one may be a major and the other a minor. There are pluses and minuses that relate to the generic interests of the student. So the reader is urged to read each chapter carefully and with due regard for content and interests. Also, there are common topics in these two fields that need to be taken into account.

About the Author

Howard Eisner spent 30 years in industry and 24 years in academia. In the former, he was a working engineer, manager, executive (ORI, Inc), and president of two high-tech companies (Intercon Systems and Atlantic Research Services). In academia, he served as a professor of engineering management and distinguished research professor in the engineering school of George Washington University (GWU). At GWU he taught courses in two departments in systems engineering, technical enterprises, project management, modulation, and noise and information theory.

He has written thirteen books that relate to engineering, management, and systems. He has also given lectures, colloquia, and presentations to professional organizations such as INCOSE (International Council on Systems Engineering), government agencies (such as the Department of Defense and NASA), and the Osher Lifelong Learning Institute (OLLI).

In 1994, he was given the outstanding achievement award from the GWU Engineering alumni).

Dr. Eisner is a life fellow of the IEEE and a fellow of INCOSE and the New York Academy of Sciences. He is also a member of Tau Beta Pi, Eta Kappa Nu, Sigma Xi, and Omega Rho, various research/honor societies. He received a BEE from the City College of New York (1957), an MS degree in electrical engineering from Columbia University (1958), and a Doctor of Science degree from GWU (1966).

Since 2013 he has served as professor emeritus of engineering management and distinguished research professor at the GWU. As such, he has continued to explore advanced topics and write about systems, engineering, and management.

Other Books from Dr. Eisner

- Advanced Algebra
- Computer-Aided Systems Engineering
- Reengineering Yourself and Your Company
- Essentials of Project and Systems Engineering Management
- Systems Engineering: Building Successful Systems
- Topics in Systems
- Thinking: A Guide to Systems Engineering Problem Solving

- Systems Architecting: Methods and Examples
- Systems Engineering: 50 Lessons Learned
- What Makes the Systems Engineer Successful?
- Problem Solving: Leaning on New Thinking Skills
- Cost-Effectiveness Analysis: A Systems Engineering Perspective
- Tomorrow's Systems Engineering: Commentaries on the Profession

Introduction

1

This book provides an overview of both systems engineering and operations research, briefly examining key features of both, some of their history, and in particular, transformations that have occurred, or are likely to occur. The author stops for a moment to also look at commonalities between the two fields. As an example, both fields, at this time, have "management" elements in common. This suggests some obvious questions: are they the same, are they different, and why?

As another example, both fields have "reliability theory: as a constituent element and the treatment in both is approximately the same. Both show how to do reliability calculations in more-or-less the same way. On the other hand, both have decision analysis as an element, but the treatment in both is invariably not the same. Reason? They come at the problem from different perspectives as well as different starting points. Also, the operations research approach tends to be more mathematical. The latter is true for several of the constructs of operations research in contrast to systems engineering. These differences are not considered negatively – they are useful in that they represent natural differences in how engineers vs. mathematicians might approach problem areas.

The high school graduate invariably is faced with the question of where to go to college and what should you select as a major and this book is designed to try to help with those questions. If the student is oriented toward analysis, the two choices might be engineering vs. analytics, the latter being offered in a business school or in the operations research department of a liberal arts college. There is quite a lot of information in this book that will guide the parent in helping his or her student make a selection between the two possibilities. Or, on the day of the double major, the student may wish to go for exactly that – a double major, with the answer being both rather than either-or. And another perspective might lie in the choice of electives, where more than one attractive elective is offered.

In the final analysis, the student may be attracted to the mathematics of operations research vs. the engineering of systems engineering. Both have

DOI: 10.1201/9781003306610-1

heavy-duty analysis foundations, but the areas of application may be quite different. Numbers on a spreadsheet related to a company's financials may be more attractive than the numbers on a graph of month-to-month projections that involve the vagaries of forecasting.

So join in the fray, make use of the information presented herein, and try to make a good choice as to which road to travel.

Overview of Operations Research

2

Operations Research is a set of analysis and/or synthesis activities leading to problem solutions. It is often an attempt to formulate a set of operations using a given (existing) set of resources in the best possible way. Thus it may be in search of an optimum procedure, or algorithm, representing a solution to a problem, or set of problems. These activities may be purely analysis, synthesis, or a combination of the two. Table 1.1 below provides a list of typical operations research domain areas.

TABLE 1.1 Illustrative Operations Research Problem area Domains

- College course registration procedures
- Purchase of goods for aircraft manufacture
- Bank services procedures
- Search for and killing of enemy submarines (ASW)
- Factory supplies and inventories
- Bomber sortie planning and execution to defeat an enemy
- Delivery of packages to sets of customers
- The "traveling salesman" problem
- Attack and defense scenarios in warfare
- Supply chain analysis and evaluation
- Flows in Networks
- Crew Scheduling
- Facility locations
- Vehicle routing

Three sets of well-known practitioners and teachers of operations research [1,2,3] have provided their definitions of operations research:

- **Kimball and Morse:** We take note of this classic treatise on Operations Research [1], as it provided support to several tasks of warfare. Topics included were as follows:

DOI: 10.1201/9781003306610-2

3

    a. Probability
    b. Measures of effectiveness (MOEs)
    c. Strategical kinematics
    d. Tactical analysis
    e. Gunnery and bombardment problems
    f. Organizational and procedural problems

- **Hillier and Lieberman:** these authors [2] have provided an extensive text in which they defined operations research as:
  - "a scientific approach to decision making that involves the operations of organizational systems"
  - "operations research involves an attempt to find the best, or optimal, solution, using scientific analyses"
- **Churchman, Ackoff, and Arnoff:** These prolific authors wrote an early book (1957) on Operations Research [3], emphasizing such topics as inventory models, allocation models, waiting time models, replacement models, and competitive models, and defining Operations Research as "an overall understanding of optimal solutions to executive-type problems in organizations"

These authors addressed the following problem areas in considerable detail:

- The transportation problem
- PERT/CPM
- The assignment problem
- School rezoning
- Inventory control
- Decision trees

They also suggested what it takes to have a career in operations research with three complementary types of academic training. The first type includes what they called the basics of operations research. This included:

- Basic methodologies of math and science
- Linear algebra
- Matrix theory
- Probability theory
- Statistical inference
- Stochastic processes
- Computer science
- Microeconomics
- Accounting

- Business administration
- Organization theory
- Behavioral science

The second type of training is in operations research itself, with topics such as:

- Linear and non-linear programming
- Dynamic programming
- Inventory theory
- Network flow theory
- Queuing models
- Reliability
- Game theory
- Simulation

Finally, as a third area of expertise, these authors recommend specialized training in some other field, for example:

- Mathematics
- Statistics
- Industrial engineering
- Business
- Economics

SOLUTION DOMAINS AND TOOLS

There are many tools that have been developed over the years in order to address operations research problems. A list of such tools and procedures is provided below in Table 1.2.

TABLE 1.2   Illustrative Tools for Solving Operations Research Problems

- Modeling and Simulation
- Probability Theory
- Statistics
- Optimization Theory
- The Simplex Method and Dual Simplex
- Simulation

(Continued)

TABLE 1.2 (Continued) Illustrative Tools for Solving Operations Research Problems

- Integer Programming
- Dynamic Programming
- Reliability Theory
- The Transportation Algorithm
- Network Analysis
- Forecasting
- Inventory Theory
- Markov Chains
- Decision Theory
- Analytics
- Calculus
- Linear and Non-linear Programming
- Convex Programming
- Duality Theory
- Stochastic analysis
- Computational techniques
- Queuing Theory
- Search Theory
- LaGrange Multiplier Procedures

The reader can see how challenging a program in operations research can be from the above list. As such, we would expect various college curricula to offer courses in these subjects, and such is generally the case (see chapter 3). Departments of operations research can "fit" in both engineering and business schools, depending upon specialty areas and often, the history and college politics. Here is a list (see below) of selected operations research departments across the United States:

Illustrative Departments of Operations Research in the United States
- Department of Operations Research and Management Science – UCA, Berkeley
- Department of Statistics and Operations Research – UNC State
- Department of Operations Research and Financial Engineering – Princeton
- Departments of Operations Research – MIT, Cornell, Stanford, Dartmouth
- Departments of Operations Research and Industrial Engineering
- Departments of Operations Research and Systems Engineering

SINGLE COURSE CURRICULUM – OPERATIONS RESEARCH

A sampling of the curriculum of operations research is provided below, from the GWU bulletin [4]

- **Operations Research Methods** – Deterministic and stochastic methods; branch and bound combinatorial algorithms; optimization algorithms; simplex method; convexity, duality; heuristic methods, sensitivity analysis; stochastic optimization; marginal analysis; Markov chains; Markov decision processes
- **Stochastic foundations of operations research** – Topics in probability theory; stochastic processes; statistical inference; foundations of probability; conditional probability and expectation; Poisson processes; Markov chains; Brownian motion
- **Mathematics in operations research** – Mathematical foundations in optimization theory; linear algebra; advanced calculus; convexity theory; geometrical interpretations and use of software
- **Topics in Optimization** – Linear, non-linear, and dynamic programming; heuristics, constraint programming; combinatorial and optimization problems; algorithms and applications; network cost flows; optimal matchings; routing problems; complexity theory; enumeration and cutting plane methods for solving integer programs
- **Applied Optimization Modeling** – Analysis of linear, non-linear, and integer models of decision problems that arise in industry; business and government; use of optimization software to solve models

MILITARY PROBLEM ADDRESSED BY OPERATIONS RESEARCH

Operations Research had many of its earliest days addressing military problems. One of the early problems arose and was considered by Lanchester who formulated his well-known Lanchester equations.

The field of operations research has its roots in "warfare" analysis and problems facing the British during WWII. If we include the Lanchester equations, we go back even to WWI. Thus it may be concluded that the UK had a lot to do with developing and promoting this field of investigation and study. It has been very successful over the years and has joined with the management folks (management science) to broaden the base and areas of interest. The importance of operations research in regard to warfare was strong enough to lead to a separate organization called MORS – the military operations research society.

MILITARY OPERATIONS RESEARCH SOCIETY

This important society has defined its vision as "becoming the recognized leader in advancing the national security analytic community through the advancement and application of the interdisciplinary field of operations research to national security issues, being responsive to our constituents, enabling collaboration and development opportunities, and expanding our membership and disciplines, while maintaining our profession's heritage".

MORS has held symposia continuously from 1957 (at the NOL – Corona) to the current time of 2020 (held at the U.S. Coast Guard). This author contributed to an early MORS conference [5] with an article on the use of Parametric Dependency Diagramming in solving military problems.

MORS has addressed a variety of military problems and issues over the years. Here is a short sample of such:

- Combat science
- Antisubmarine warfare (ASW)
- Submarine hiding strategies
- Troop defense deployments
- Troop attack scenarios
- Mine laying
- Bombing sortie strategies
- Fighter aircraft strategies (energy theory)
- The Lanchester equations

Additional information regarding the last item on the above list follows.

THE LANCHESTER EQUATIONS [6]

A good example of a quantitative approach to warfare can be seen in the Lanchester equations. They address the relative strength of opposing military forces using a set of differential equations. There are two laws: the linear law (for ancient warfare) and the square law (for more modern warfare). For the linear law, we have two armies, each shooting at one another with a stream of bullets. One side (A) has offensive power α which is the number of soldiers it can capacitate per unit time. Similarly for side B, with offensive power β.

One form of the Lanchester equations is simply:

$$dA/dt = -\beta B \quad \text{and} \quad dB/dt = -\alpha A$$

Solutions and commentaries then follow in terms of losses by each side as a function of the time variable.

Apparently, studies by zoologists have found that chimpanzees intuitively follow the Lanchester square law in that they will not attack another group unless the advantage is at least a factor of 1.5. Similar projects have analyzed Pickett's charge of confederate infantry against the Union infantry during the 1863 Battle of Gettysburg and also battles between British and German air forces in 1940, both using Lanchester equations.

SOME FAMILIAR NAMES IN OPERATIONS RESEARCH

Over the years, as problem solutions have been found, the names of the problem solver have come along with the method. The following is a short list of some of these pioneers in the field.

Lanchester – Lanchester's Laws and equations
Dantzig – The simplex method
Karmarkar – the Karmarkar algorithm
Chapman and Kolmogorov – the Chapman-Kolmogorov equations
Markov – Markovian decision models
Forrester – Forrester's Urban/industrial dynamics

Kuhn and Tucker – the Kuhn-Tucker conditions
Oscar Morgenstern and von Neumann– Game Theory
Additional History

As suggested earlier in this chapter, much of the history of operations research can be found with the British both before and during WWII. There was a tendency in the UK to call the field "operational research", with sub-disciplines identified as follows [7]:

- Financial engineering
- Computing and information technologies
- Policy modeling and public sector work
- Revenue management
- Simulation
- Stochastic models
- Transportation
- Manufacturing, service sciences, and supply chain management

The British are well-known for operations analysis of radar systems as well as the operations of convoys and bombers under a variety of warfare circumstances.

THE OPERATION RESEARCH SOCIETY OF AMERICA (ORSA) [6]

Apparently, this society was born in Harriman, New York on May 26, 1952, with Philip Morse as the first President. This society started with a strong emphasis on military matters, with opposing thoughts and emphasis on management issues. The latter led to the creation of The Institute of Management Science, with William Cooper as the first TIMS president. A competition of ideas ensued over the next several years until a merger between ORA and TIMS occurred. This was a stable configuration, with a softening of the differences between the two organizations and ideas. Today, if you will, there is a wide range of acceptable topics, with management contributions taking their appropriate place.

A relatively recent investigation of the history of operations research [7] from William Thomas suggests that there is much to be learned from a deep look at the early history of operations research and related fields. This author basically claimed that operations research had a massive and deep effect on policy in two countries over a 20-year period. That point of view is accepted here and dealt with in the treatment from the several chapters.

One such example was the establishment of the Army's Operation Research Office (ORO), headed originally by Ellis Johnson. That activity reached out to the White House in order to suggest a National Research Council committee to explore "national political objectives". These types of ambitious thinking and actions were reflected in the "Science of Better", as well as "Systems Thinking" and "soft O.R." All of this set the stage for strong pushes in the direction of the significance of O.R., espoused by many though not without several detractors over the years.

William Thomas's history argues that O.R. assumed "the mantle of a profession" by the mid-1950s, as seen by such topics as linear programming, inventory theory, search theory, and queuing theory. These had lives of their own as they emerged from associations with the margins of mathematics, statistics, and economics. "Math Modeling" was on the team of advances and led the charge for quite a few years, along with practitioners (as professors) who made contributions in the literature as well as for real-world customers that they were serving.

REFERENCES

1. Kimball, G., and P. Morse, "Methods of Operations Research", Library of Congress, 1946
2. Hillier, F., and G. Lieberman, "Introduction to Operations Research", 3rd Edition, Holden-Day Inc., 1980
3. Churchman, C.W., R.L. Ackoff, and E. Leonard Arnoff, "Introduction to Operations Research", John Wiley, 1957
4. GWU Bulletin, Operations Research Courses
5. Eisner, H., "The Use of Parameter Dependency Diagramming in Military Problem Solving" MORS Meeting, 1967, National Bureau of Standards
6. Lanchester's Equations, Wikipedia
7. ORSA history article, Thomas, W., "Rational Action: the sciences of policy in Britain and America 1940–1960"

Overview of Systems Engineering

3

The field of systems engineering addresses the best way of building systems. This author's definition, provided in the INCOSE Handbook [1], is:

> "Systems Engineering is an iterative process of top-down synthesis, development, and operation of a real-world system that satisfies, in a near optimal manner, the full range of requirements for the system"

That same handbook defines the elements of systems engineering, taking off from the 15288 standard [2], as a set of processes, namely:

1. Technical Processes
   - Business or mission analysis process
   - Stakeholder needs and requirements definition process
   - System requirements definition process
   - Architecture definition process
   - Design definition process
   - System analysis process
   - Implementation process
   - Integration process
   - Verification process
   - Transition process
   - Validation process
   - Operation process
   - Maintenance process
   - Disposal process
2. Technical Management Processes
   - Project planning process
   - Decision management process
   - Risk management process

DOI: 10.1201/9781003306610-3

- Configuration management process
- Information management process
- Measurement process
- Quality assurance process
3. Agreement Processes
    - Acquisition process
    - Supply process
4. Organizational Project-enabling Processes
    - Life cycle model management process
    - Infrastructure management process
    - Portfolio management process
    - Human resource management process
    - Quality management process
    - Knowledge management process

A senior systems engineer, Dennis Buede [3], has pointed out that systems engineering goes back to the early '40s and even the 1900s at Bell Labs, and the first course on this topic at MIT, taught by Mr. Gilman, the Director of systems engineering at Bell Labs.

Yet another way to explore the scope of systems engineering is to examine the 30 elements of systems engineering, cited in this author's early book on systems engineering management [4], as follows:

1. Needs/goals/objectives
2. Mission engineering
3. Requirements analysis/allocation
4. Functional analysis/decomposition
5. Architecture design/synthesis
6. Alternatives analysis/evaluation
7. Technical performance measurement (TPM)
8. Life cycle costing
9. Risk analysis
10. Concurrent engineering
11. Specification development
12. Hardware/software/human engineering
13. Interface control
14. Computer tool evaluation and utilization
15. Technical data management and documentation
16. Integrated logistics support (ILS)
17. Reliability, availability, maintainability (RAM)
18. Integration
19. Verification and validation

20. Test and evaluation
21. Quality assurance and management
22. Configuration management
23. Specialty engineering
24. Replanned product improvement (P3I)
25. Training
26. Production and Deployment
27. Operations and Maintenance (O & M)
28. Operations evaluation/reengineering
29. Systems disposal
30. Systems engineering management

Another way to view systems engineering is from the Body of Knowledge perspective as developed by the SERC some years ago [5]. The knowledge areas are delineated below. We note that this approach cites many different aspects of systems engineering, as compared with for example the 15288 standard:

1. Introduction to SEBoK
2. Introduction to Systems Engineering
3. SEBoK Users and Uses
4. Systems Fundamentals
5. Systems Approach Applied to Engineered Systems
6. Systems Science
7. Systems Thinking
8. Representing Systems with Models
9. Systems Engineering and Management
10. Introduction to Life Cycle Processes
11. Life Cycle Models
12. Concept Definition
13. System Definition
14. System Realization
15. System Deployment and Use
16. Systems Engineering Management
17. Product and Service Life Management
18. Systems Engineering Standards
19. Product Systems Engineering
20. Service Systems Engineering
21. Enterprise Systems Engineering
22. System of Systems
23. Healthcare Systems Engineering
24. Enabling Systems and Enterprises
25. Enabling Teams
26. Enabling Individuals

Although there is considerable agreement as to the elements of systems engineering, there is not always the agreement that these elements are being executed appropriately. An ex-administrator of NASA, Michael Griffin, has basically declared that systems engineering needs fixing [6]. He suggested that systems engineering should lead to products (systems) that are elegant, and that a variety of experiments need to be run to find a process with that kind of result.

DEFINING "SYSTEM"

We often take it for granted that we know precisely what is meant by a "system" that is part of systems engineering. Several INCOSE Fellows have decided to tackle this "problem", and have documented a comprehensive approach to defining a system [1, p. 20]. These authors claim that a well-conceived definition should communicate the meaning of system more effectively across communities of research and practice. It should also learn and adopt techniques from other communities and it should improve various communities' understanding of the definition of a system. These researchers have therefore proposed a family of definitions, related to the common theme of emergence, which is in line with the realist and constructivist worldviews and better reflects the current scope of systems engineering.

A HISTORICAL PERSPECTIVE

The INCOSE Handbook has documented some of the important dates in the history of systems engineering as a discipline. These dates are reiterated as follows:

- 1937: British team to analyze the air defense system
- 939–1945: Bell Labs supporting NIKE program
- 1951–1980: SAGE air defense system, supported by MIT
- 1954: RAND Corporation suggests term "systems engineering
- 1956: RAND defines systems analysis
- 1962: Publication of A. D. Hall's book on systems engineering (see below)
- 1969: modeling of systems from Jay Forrester

- 1990: NCOSE established
- 1995: NCOSE expanded to incorporate international people and views
- 2008: Issuance of standard 15288

An earlier purveyor of the elements of systems engineering was A.D. Hall from Bell Labs [7]. His exploration included:

a. The five phases of systems engineering
b. Problem-solving models
c. Systems synthesis
d. Systems analysis
e. Selecting an optimum system
f. Communication problems
g. The role of measurement
h. The economic theory of value
i. The psychological theory of value
j. Statistical decision-making and games
k. Graphical models
l. Input-output and functional design
m. Psychological aspects of synthesis

Another source of basic information as to the nature of systems engineering can be found in the classic treatise from Goode and Machol [8]. The Table of Contents reveals what the authors considered to be the elements of systems engineering. A sampling of some of these topics includes:

a. Distributions of discrete and continuous variables
b. Statistics
c. Mathematical models
d. Design and analysis of experiments
e. Exterior system design
f. Tools of interior system design
g. Digital and analog computers
h. System logic
i. Queuing theory
j. Game theory
k. Linear programming
l. Group dynamics
m. Cybernetics
n. Information theory
o. Servomechanism theory
p. Human engineering

In the same vein, we look at Machol's systems engineering handbook [9] to find out more about what that early researcher had to say about this important but somewhat elusive topic. Here are some entries from his handbook:

a. System environments
b. System components
c. System theory
d. System techniques
e. Useful mathematics

The section on system theory has the following breakdown:

a. Information theory
b. Game theory
c. Decision theory
d. The simplex method
e. Linear programming
f. Dynamic programming
g. Queues and Markov processes
h. Feedback theory
i. Control systems

Thus we see many interpretations of the content of systems engineering from about the end of WWII to the present time. What are we to make of these disparate interpretations?

The answer is actually easy to come by. The evolution of systems engineering was a natural process, leading to the more established and accepted definition espoused by INCOSE, the international council on systems engineering. Thus we look to the handbook and also to the 15288 standard as up-to-date definitions.

A BUMP IN THE ROAD

In addition to Griffin's questioning of "how to fix" systems engineering, we have a well-known author and INCOSE Fellow posing what he calls the systems engineering conundrum – where is the engineering? [10]. Wasson's main point is that INCOSE treats systems engineering more as a management activity rather than an engineering one. He goes on to try to prove the point with lots of examples and an exposition of what he calls the "technical competency void". This author tends to accept much of Wasson's argument and is looking for a general reaction among the INCOSE members and fellows. This is an issue that is likely to rise again, with a considerable amount of activity and discussion.

A SIGNIFICANT MILESTONE AND CONTRIBUTION

Eberhardt Rechtin, a master systems engineer, wrote the seminal treatise regarding systems architecting [11]. In that book, he took considerable time to set forth his experiences with building systems in the form of a series of "heuristics". A sample of such is provided here in the list below:

-"Simplify, simplify, simplify

-The greatest leverage in systems architecting is at the interfaces

-In architecting a software system, all the mistakes are made in the first day

-Use analogies and metaphors

-A model is not reality

-If it ain't broke, don't fix it

-Quality cannot be tested in, it must be built in

-The first look analyses are often wrong

-No complex system can be optimum for all parties concerned

-80% of the useful work in an organization is accomplished by 20% of the people

FIGURE 3.1   Sample of Heuristics provided by E. Rechtin [11].

EVOLUTION OF SYSTEMS ENGINEERING IN THE DEPARTMENT OF DEFENSE

The DoD has continually supported the study and use of systems engineering over the years. One very significant way in which they have done that is to establish systems engineering research centers (SERCs) and supported them for some number of years. In addition, the DoD has provided guidance to the community at large, and directly in terms of talks at meetings and conferences.

As an example, a major in the Air Force has documented [12] the evolution of systems engineering in the DoD. Here are some of the key points that he has made in doing so:

- Systems engineering started in the early '40 s at Bell Labs
- RAND defined some early aspects of systems engineering, noted by the War department
- RAND continued to define systems principles in regard to key physical elements of systems such as:
  1. Aircraft (manned and unmanned)
  2. Ships
  3. Datalinks
  4. Airborne radars
  5. Command centers
  6. Satellites
  7. Ground controller
- A PPBS (planning, programming, and budgeting system)
- Lessons from failures and breaches
- A Systems Engineering Plan (SEP) from each of the services
- Eight Core areas for better buying power include:
  1. Achieving affordable goals
  2. Achieving dominant capabilities and controlled life cycle costs
  3. Incentivizing industry toward improved productivity
  4. Incentivizing innovation
  5. Eliminating non-productive activities and processes
  6. Promoting effective competition
  7. Improving service acquisitions
  8. Improving professionalism in the overall workforce
  9. Improving all aspects of key acquisition programs

In addition to Major Page's presentation, as above, Ms. Kristen Baldwin in the DoD has provided a systems engineering update in the DoD [13], addressing the following focus areas:

- Growing engineering and technical leadership talent
- Bringing modularity, agility, and innovation into systems
- Improving digital acquisition, engineering, and manufacturing practices
- Addressing complex software development challenges
- Establishing cyber-resilient aerospace and defense systems
- Enabling trust and access to assured hardware and software
- Improving enterprise and mission integration management capabilities

- The following concrete deliverables:
  1. Digital engineering
  2. The modular open systems approach (MOSA)
  3. Software
  4. Mission and SoS engineering
  5. Workforce progress
  6. Master schedule
  7. Sustainment engineering

Ms. Baldwin also identifies opportunities for collaboration between DoD and Industry as:

  a. Digital engineering
  b. Modeling and simulation
  c. The Modular Open Systems Approach (MOSA)
  d. Reliability and Maintainability (R & M) in Weapon System Design
  e. Sustainment Engineering
  f. System of Systems/Mission Engineering
  g. Workforce
  h. Cyber Resilient Weapon Systems
  i. Systems Engineering Research

LOOKING TO THE FUTURE

INCOSE has taken the time to explore what might lie ahead for the field of systems engineering [14] by posing and answering the following five questions in relation to the next 15 years (by 2035):

1. What are the changing societal needs that systems engineering can and should respond to?
2. What are the technology trends that systems engineering can incorporate?
3. What are the systems engineering changes that will be responsive to (1) and (2) above?
4. What is the likely effect of digital engineering on systems engineering?
5. How will and should systems engineering respond to needs in cyber-security, trust, resilience, and ease of use?

THE GRAND UNIFIED THEORY OF SYSTEMS ENGINEERING [15]

It has been suggested, over the years, that the field of systems engineering would benefit from a grand unified theory. Dr. J. Kasser and Y-Y Zhao have stepped up to this challenge with a paper called "Towards a Grand Unified Theory of Systems Engineering (GUTSE)". The essence of their approach is to define some seven elements that could form the basis for a GUTSE. These elements are as follows:

1. Holistic thinking
2. A Hitchins-Kasser-Massie framework (HKMF) for understanding systems engineering
3. Distinguishing between the role of systems engineering (SETR) and the activity (SETA)
4. Pure, Applied, and Domain systems engineering
5. A problem classification matrix
6. A systems engineering competency maturity model framework
7. The nine-systems model

REFERENCES

1. Walden, D., et. al "INCOSE Systems Engineering Handbook", 4th edition, John Wiley, 2015
2. Standard 15288, "System Life Cycle Processes"
3. Buede, D., "History of Systems Engineering", see https://www.incos.org/about-systems-engineering/historyofsystemsengineering
4. Eisner, H., "Essentials of Project and Systems Engineering Management", John Wiley
5. SEBoK, from the SERC (Systems Engineering Research Center, Stevens Institute of Technology, Hoboken, New Jersey
6. Griffin, M., "Michael Griffin Explains How To Fix Systems Engineering", Stevens Institute of Technology, talk, 14 December 2010
7. Hall, A.D., "A Methodology for Systems Engineering", Van Nostrand, 1962
8. Goode, H. and R. Machol, "Systems Engineering", McGraw-Hill, 1957
9. Machol, R., "Systems Engineering Handbook", McGraw-Hill, 1965
10. Wasson, C., "The Systems Engineering Conundrum", INCOSE International Symposium, Volume 31, Issue 1, September 2021

11. Rechtin, E., "Systems Architecting", Prentice-Hall, 1991
12. Page, Austin, see https://SdmMIT.du/News/TheEvolutionofSystemsEngineeringin theDOD
13. Baldwin, K., "DoD Systems Engineering Update"
14. INCOSE Systems Engineering Vision 2035
15. Kasser, J., and Y-Y Zhao, "Towards a Grand Unified Theory of Systems Engineering", SETE 2014

Academic Representations

4

Our colleges and universities tend to be leaders in both operations research and systems engineering. That is, they are "run" by professors, and these professors are at the leading edge of research in both operations research and systems engineering. So, in this chapter, we will take a broad look at curricula from various well-known universities, including:

a. Johns Hopkins
b. MIT
c. Stevens Institute of Technology
d. The George Washington University
e. Michigan
f. University of Maryland Baltimore Campus (UMBC)
g. Stanford University
h. California Institute of Technology (CalTech)
i. UC San Diego
j. London School of Economics, UK
k. UC San Diego
l. London School of Economics
m. George Mason University
n. Embry Riddle
o. Boston University

JOHNS HOPKINS ENGINEERING

Johns Hopkins has a part-time master's program that they claim will "help you build skills" that translate to the work environment. Hopkins engineering courses are designed for online delivery at a pace that suits your life with practical knowledge that elevates your career. Engineers are needed now more than ever – A degree from Johns Hopkins can open doors for you.

DOI: 10.1201/9781003306610-4

Johns Hopkins also offers an MS in Operations Research with the following specific courses:

- Nonlinear Optimization I
- Nonlinear Optimization II
- Combinatorial Optimization
- Risk and Decision Analysis
- Uncertainty Modeling
- Network Models
- Stochastic Processes
- Time Series Analysis
- Economic Foundations for Public Decision Making
- Monte Carlo Methods
- Quantitative Portfolio Theory and Performance Analysis
- Urban and Environmental Systems
- Mathematical Modeling and Consulting

You will have earned a master's degree after just 10 courses. Johns Hopkins has a top/-ranked systems thinking program. They claim that U.S. News and World report lists their systems engineering program as # 3 nationally. They also claim that "we sharpen your ability to discover solutions, manage projects, and monitor performance, learn to think more deeply, and solve smarter through modern, relevant courses". Choose from six focus areas: systems, cybersecurity, human systems, modeling and simulation, project management, and software systems.

Hopkins also suggests that systems engineers are needed for a broad range of work. What you learn applies to just about any industry, including civil engineering, medicine, robotics, transportation, and aerospace. Students have explored systems to improve forklift safety, develop autonomous vehicles, design aviator tracking systems, and structure personal air vehicles.

MIT

MIT is acknowledged to be one of the preeminent engineering schools in the country. As such, it offers a System Design and Management program, which bears a definitive relationship to systems engineering. The SDM program emphasizes systems thinking. The MIT bulletin sets forth the following propositions with respect to SDM:

"Over the last one hundred years, engineering has evolved dramatically from a singular focus on emerging technology and infrastructures to a broader view of these elements as dynamic, interconnected systems. Systems design and Management asks individuals to anticipate and resolve challenges before they arise. Leaders in the field are required to think holistically, navigating constant uncertainty with a keen sense of flexibility".

"Systems Design and Management draws upon advanced engineering and management principles, methods and tools to address the world's most daunting challenges. Our students build upon real-world experience in their field, gaining skills and perspective through rigorous academic courses, research, and hands-on experience across MIT".

STEVENS INSTITUTE OF TECHNOLOGY

Stevens Offers a Master's of Engineering in Systems Engineering.

The online application requires a $60 nonrefundable fee, two letters of recommendation, official transcripts from all institutions attended, and a GRE test score. Core requirements are listed below:

Modeling, Simulation and Analysis – Systems Modeling and Simulation

Management – Project Management of Complex Systems

Concept – SYS 625 Fundamentals of Systems Engineering

For Architecture and Design – SYS 650 System Architecture and Design; SYS 672 – Design of CPS: Ensuring Systems Work and Are Robust; SYS 565 – Software Architecture and Component-Based Design

For Implementation – SYS 605 – Systems Integration; SYS 67 – Implementation of CPS: Bringing Solutions to Life; SYS 567 – Software Testing, Quality Assurance and Maintenance

For Sustainment – SYS 640 – Systems Supportability and Logistics; SYS 674 – Sustainment of CPS: Managing Evolution; Cost-Effective Space Mission Operations

THE GEORGE WASHINGTON UNIVERSITY

GWU offers a Master's in Systems Engineering with a curriculum that features the following topics:

- Project management
- Accounting and finance
- Risk management
- Systems analysis and evaluation
- Architecture analysis and advanced modeling
- Axiomatic design
- Requirements management
- Quality assessment
- Large-scale systems

Specific core courses are:

- The Management of Technical Organizations
- Decision Making With Uncertainty   ·
- Problems of Engineering Management and Systems Engineering
- Survey of Finance and Engineering Economics
- Systems Engineering
- Program and Project Management

Required courses for an MS in Systems Engineering are:

- Systems Engineering II
- Systems Analysis and Management
- Requirements Engineering
- Model-Based Systems Engineering
- Project Cost and Quality Management
- Applied Enterprise Systems Engineering
- Software Systems Engineering

UNIVERSITY OF MICHIGAN

The University of Michigan offers a master's degree which they call "systems engineering and design". This program requires 30 total credits and is offered online.

Breadth courses are:

- Introduction to systems engineering
- Development and verification of system design
- System architecting concept development and embodiment

A systems elective can be chosen from among the following:

- Risk analysis
- Design for six sigma
- Software systems engineering

A design-focused elective can be chosen from among the following:

- Space systems design
- Multidisciplinary design optimization
- Design of environmental engineering systems
- Analytical product design
- Design of digital control systems
- Advanced design for manufacturability
- Design optimization
- Marine design
- Nuclear core design

Other elective specialties are as follows:

- Automotive engineering
- Manufacturing
- Energy systems engineering
- Aerospace engineering

There are downloadable plans of study for the online student and the on-campus student.

THE UNIVERSITY OF MARYLAND AT BALTIMORE CAMPUS

The University of Maryland at Baltimore Campus (UMBC) offers an MS in systems engineering as well as graduate certificate programs that accelerate the development of systems engineers by providing practical, real-world experiences that can be immediately applied on the job. The on-campus location is in Catonsville, Maryland.

The specific MS program takes three years on a part-time basis, requiring 0 credits. The cost is 2 per credit hour for resident students. Three credits are selected from the three courses:

- Risk Analysis
- Design for Six Sigma
- Software Systems Engineering

A primary program at UMBC is called systems engineering and PLM: time to converge 20 years of product development processes and learnings.
Further emphasis is placed upon:

- Digital transformation
- Digital thread
- The resilient platform
- Systems Engineering principles
- System Architecture and Design
- System Implementation, Integration, and Test
- Modeling, Simulation, and Analysis
- Decision and Risk Analysis

STANFORD UNIVERSITY

Stanford is known as one of our leading Universities in a wide variety of subject areas. A key designated program is described as Management Science and engineering. The Management Science focus can be considered to be related to operations Research and the Engineering focus more-so on systems engineering. Their MS program requires a minimum of 45 units above and beyond a BS degree. All students are required to develop competence, especially in optimization and analytics, organizations and decisions, and probability. Specialties are available in seven areas, as follows:

1. Financial Analytics
2. Operations and Analytics
3. Technology and Engineering Management
4. Computational Social Science
5. Decision and Risk Analysis
6. Energy and Environment
7. Health Systems Modeling

The student addresses one or more of these specialty areas that carry out assignments in specific courses and problem identification and solving. High standards and practical issue analysis are embedded in these courses, with specific written assignments every week.

CALTECH (CALIFORNIA INSTITUTE OF TECHNOLOGY)

CalTech offers an MS in Model-Based Systems Engineering. It claims that the student will learn to:

* Scope and execute full lifecycle MBSE tasks
* Investigate internal and external interactions
* Evaluate, critique, and improve SysML structures and models
* Craft key representation, diagrams, and use-cases

MBSE is considered to be a major foundation of the study of systems engineering and has remained so for a couple of decades, under the current leadership of Sandy Friedenthal and the original idea of Prof. Wayne Wymore at Arizona State.

UC SAN DIEGO

This University offers a master's degree in systems engineering featuring the following required courses:

* Systems Engineering Management
* Systems Requirements Analysis
* Systems Verification and Validation
* Systems Engineering Software Overview
* Systems Hardware/Software Integration

LSE IN THE UK; MS IN OPERATIONS RESEARCH AND ANALYTICS

The LSE (London School of Economics) offers a degree and study program in Operations Research and Analytics and is a good example of adding analytics to operations research and an example of a transformation of the field itself. Courses include:

- Management and Strategy
- Financial Statistics
- Operations Research
- Analytics

It is claimed that there are four types of analytics:

1. Descriptive (What have we done in the past?)
2. Diagnostics (Why have we seen past results?)
3. Predictive (Where are we going and when?)
4. Prescriptive (How should we take action?)

A reasonable definition of analytics can be articulated as:

> The analysis of data, typically large sets of business data, using mathematics, statistics and computer software; patterns and meaningful information gathered from the analysis of data.

GEORGE MASON

This master's degree is ideal for engineers who strive to attain technical leadership positions in their organizations. Addressing a broad range of issues relevant to the design, implementation, analysis, and management of systems, this program prepares students for a professional career working with large complex systems of the future. Such systems include:

- The national air transportation system
- Computer networks
- Autonomous vehicles
- Intelligent robots
- The electric power grid
- Healthcare systems
- Financial trading systems

The cost per credit hour for VA state residents is 813.25 and for out-of-state is 1581.25.

Students in the program benefit from a mature corporate relationship with the Volgenau School of Engineering.

Areas of specialization in Operations Research include:

a. Optimization
b. Stochastic modeling
c. Predictive data analytics
d. Operational analysis problems
e. Decision analysis
f. Military operation research

EMBRY RIDDLE AERONAUTICAL UNIVERSITY

Embry Riddle offers the MS in Systems engineering and specializes in students in one or other fields within the aeronautical field. They address real-world issues pertaining to transportation and aeronautics. They are oriented toward the working professional.

BOSTON UNIVERSITY

Boston University offers an MS in Systems Engineering with three courses as the core, two courses from the concentration, and 4–8 credits as electives. Among the selection of core courses are as follows:

1. Dynamic Systems Theory
2. Optimization Theory and Methods
3. Advanced Stochastic Modeling and Simulation

The possible selection for the concentration are as follows:

1. Computational & Systems Biology (8 courses)
2. Control Systems (12 courses)
3. Energy and Environmental Systems (10 courses)
4. Network Systems (7 courses)
5. Operations Research(13 courses)
6. Production and Service Systems (8 Courses)

Suggested electives have an array of 12 courses that might be selected by the student.

REFERENCE

See the list of colleges and universities at beginning of chapter three

Systems Engineering: Revisited

5

There are many who have seen the beginnings of systems engineering with A.D. Hall, at Bell Labs. To be more particular, there's the claim that systems engineering really had its formal beginning with a class taught at MIT by a gentleman by the name of Gilman, who was the Director of Systems engineering at Bell Labs. So it seems that the folks at Bell Labs planted the flag in the field of systems engineering before just about everyone else.

A.D. Hall defined systems engineering as a function with five phases:

1. System studies or program planning
2. Exploratory planning, which included problem definition, selecting objectives
3. Systems synthesis, systems analysis, selecting the best system, and communicating the results
4. Studies during development, to include integrating and testing important parts of the system
5. Current engineering, which involves engineering during operations and maintenance

However, there are those who claim that systems engineering began with the RAND Corporation and their definition of systems analysis. On the other hand, many would claim that systems analysis and systems engineering are far from the same and that companies like TRW represented the beginnings of systems engineering as they assisted the Air Force with its aircraft and missile systems programs. This author comes out on the TRW et al. side of that argument, recalling that in most reckonings, systems analysis is a subset of systems engineering. Today's systems engineer would very likely agree that a very good definition of systems engineering resides neatly in the excellent text of the INCOSE Systems Engineering Handbook, Fourth Edition [1]. And that Handbook is definitely based upon Standard 15288 – Systems and

Software Engineering [1] – System Life Cycle Processes. Having said that, we will recognize the validity of an INCOSE Fellow query [2] who asked the question "where is the engineering?" We will have more to say about that question, and a possible answer, later on in this book. The other attitude that is likely to be expressed by today's systems engineer is – systems engineering is whatever INCOSE declares it to be. But since systems engineers are likely to be continually questioning, on a variety of subjects, we wish to acknowledge a paper from serious INCOSE fellows [3] – they tried to answer the question – what is a system? They are both forgiven and appreciated for their curiosity and their interest in going back to fundamentals. That's the nature of systems engineering and indeed of systems engineers.

So I agree with the INCOSE Systems Engineering Handbook – that systems engineering is whatever INCOSE declares it to be. Having said that, we will respect INCOSE Wasson's query –where is the engineering? He was asking the question – is systems engineering more about management or is it, in the final analysis, more about engineering? These are important questions for today, in the systems engineering community. The issue for today may well be – will INCOSE address this question, and in what form, if they do.

Another way to come at this issue is to look at what INCOSE designates as systems engineering in their certifications. INCOSE is very clear about this matter, for obvious reasons. A simplistic view of "what is systems engineering" is that it's the set of all engineering tasks necessary to build successful systems. Before Standard 15288, we would like to go back to this author's definition from his book on "building successful systems" [4]. That source looks at the following activities as representing the systems approach:

1. Establish and follow a systematic and repeatable process
2. Assure interoperability and harmonious system operation
3. Be dedicated to the consideration of alternatives
4. Use iterations to refine and converge
5. Create a robust and slow-die system
6. Satisfy all agreed-upon user/customer requirements
7. Provide a cost-effective solution
8. Assure the system's sustainability
9. Utilize advanced technology, at appropriate levels of risk
10. Employ "systems thinking" [5]

There is no better source for the meaning of the last item on this list than Peter Senge's "The Fifth Discipline".

So this author winds up referring all readers to the INCOSE Systems Engineering Handbook, Fourth Edition, when looking at and for "Systems

Engineering Today" And as far as the work on defining a [3] system, this author commends it as interesting. This appears to be especially true if one accepts the "theme of emergence" as covering all "real and conceptual systems?"

Coming at this issue from another perspective, Watson and Griffin [6,7] have pointed at systems engineering as "the discipline of engineering elegance". They define elegance in terms of four features:

1. Efficacy
2. Robustness
3. Efficiency and
4. Unintended consequences

We are moving in the right direction if we recognize that elegance is simplicity. Watson and Griffin go on to explore the meaning of that word, by means of the following citations:

- Simplicity is not just minimalism in the absence of clutter, it involves digging through the depth of complexity; to be truly simple, you have to go deep
- They quote Einstein as saying "everything should be made as simple as possible, but not simpler"

Yet another representation of what they meant by an elegant design is based upon these four features:

1. Utility
2. Robust
3. Functional unification
4. Context

So the Watson and Griffin approach to an elegant design gives us several points of departure for yet another theory of systems engineering if we are of a mind and inclination to do so.

ARCHITECTING

The DOD approach to architecting, designated as DoDAF, remains unacceptable, and there is no sign of progress in the DoD in that respect. This author suggests that the approach in his 2019 book [8] is a good candidate for a procedure that is more than acceptable for a new and definitive process of architecting. It may also be the direction for coming up with an "elegant" architecture, following the Watson and Griffin perspectives in that direction.

Nothing is more important, among the tasks of the systems engineer, than how to architect a system. This author's approach brings together a sensible and cost-effective perspective to the process of architecting.

FAILURE IN SYSTEMS ENGINEERING

We often take it for granted that systems engineering, as herein defined, is the beginning and end of all matters having to do with building successful systems. Being particularly thoughtful, a panel of senior engineers got together to discuss and explore what could be learned about our failures in systems engineering [9]. This panel explored how systems engineers respond to and learn from failure and point in the direction of avoiding failure in the future. Each panelist was asked to provide an opening statement. These statements are summarized below and represent responses to the questions:

Provide an opening statement and discuss how systems engineers respond to and learn from failures and identify future directions important to the systems engineering community.

Gary Payton, retired Deputy Under Secretary of the Air Force for Space Programs: Mr. Payton developed lessons learned and pointed to the X-33 reusable launch Vehicle, which experience several failures during test. Payton said that the team needed to reach further to explore the failures and develop a series of lessons learned.

John Thomas, President-elect of INCOSE: Mr. Thomas explored the hypothesis: Can system failures be attributed to the systems engineer who then has the ability to assess and re-assess, and thereby prevent future failures?

Michael Griffin, former NASA Administrator: Dr. Grifffin gave examples of failure in two programs: the Hubble Space Telescope and also the first powered space interceptor, the Delta 180. The problem discovered after an interceptor kill was a small amount of hysteresis that resulted in purposing and an error of several feet in the ultimate location of the intercept. This was fixed in the next flight, apparently.

Ronald Kadish, from Booz-Allen Hamilton and former director of the Missile Defense Agency: Kadish discussed the phrase "failure is not an option", made famous from the 1995 film Apollo thirteen. A distinction was made between failures at intermediate steps, except for the final step which was not a failure. In short, the panel discussants explored the various meanings of failure, including the well-known phrase from Apollo 13, "Failure is not an option".

Dan Goldin, the former NASA administrator, had several interesting observations about his NASA job, back in April 1992[10]: Goldin served the longest as NASA Administrator (April 1992 to November 2001) and presented a quite interesting lecture, the essence of which went back to Michaelangelo and how he was able to produce what he did. In particular, Goldin referred back to a principle of design involving visualizing the solution and then working backward from there. This anecdote captured the imagination of the audience, as did Goldin's overall notion regarding systems engineering. Goldin referred to a comment from Michaelangelo – "How do I carve these beautiful structures? – I see the angel in the marble, and I carve until I set it free". It's a beautiful image but stops short of explaining how this process is carried out in the real world with a team of systems engineers. This anecdote reminded this author of Einstein's approach to finding new solutions, which were abundant to Einstein, but not to others: He tried to visualize solutions as a means of problem-solving.

CHARLES WASSON – RE-VISITED [2]

Fellow Charles Wasson stepped up to call it the conundrum – where is the engineering? Wasson believes that the INCOSE approach to systems engineering lacks a central core of engineering analysis and evaluation. This core defines its concepts, principles, and practices and is focused on management rather than engineering. Can this author think of tasks that would (or should) be added to the definition of systems engineering, as defined in the standard 15288? Certainly, that is the case. Here are seven task areas that are engineering-oriented in distinction to management:

A. Trade-off studies
B. Definition of alternatives in design
C. Interoperability analysis
D. New and specific methods of architecting
E. Cost-effectiveness analysis
F. Requirements satisfaction analysis
G. System performance modeling and simulation (M & S)
H. Single-point (mission) failure analysis

MITRE'S SYSTEMS ENGINEERING

MITRE, a leading company and FFRDC in service of the Air Force and other government agencies, has taken the time to elaborate on their approach to systems [11]. Their SEG – Systems Engineering Guide has the following topics, each covered in some detail:

- Enterprise Engineering
- Enterprise Planning and Management
- Enterprise Technology, Information, and Infrastructure
- Engineering Information- Intensive Enterprises
- Systems of Systems
- Systems Engineering for Mission Assurance
- Transformation Planning and Organizational Change
- Enterprise Governance
- MITRE FFRDC Independent Assessment
- Life Cycle Building Blocks
- Concept Development
- Requirements Engineering
- System Architecture
- System Design and Development
- Systems Integration
- Test and Evaluation
- Implementation, O & M and Transition
- Acquisition Systems Engineering
- Risk Management
- Configuration Management
- Integrated Logistics Support
- Quality Assurance and Measurement
- Continuous Process Improvement

It is noted that the MITRE version of systems engineering includes TQM concepts and several management tasks.

NASA'S VIEW OF SYSTEMS ENGINEERING

NASA has definitely kept up with systems engineering as a core discipline with respect to their building successful systems. They have produced and followed a systems engineering handbook [12], with a table of contents as described as follows:

2.1. Fundamentals of Systems Engineering
2.2. The Common Technical Processes and the SE Engine
2.3. An Overview of the SE Engine by Project Phase
2.4. Example of Using the SE Engine
2.5. Distinctions Between Product Verification and Product Validation
2.6. Cost-Effectiveness Considerations
2.7. Human Systems Integration (HIS) in the SE Process
2.8. Competency Model for Systems Engineers

3.1. NASA Program/Project Life Cycle
3.2. Program Formulation
3.3. Program Implementation
3.4. Project Pre-Phase A Concept Studies
3.5. Project Pre-Phase A Concept and Technology Devekopment
3.6. Project Phase B Preliminary Design and Technology Completion
3.7. Project Phase C: Final Design and Fabrication
3.8. Project Phase D: System Assembly, Integration and Test, Launch
3.9. Project Phase E: Operations and Sustainment
3.10. Project Phase F: Closeout
3.11. Funding, the Budget Cycle
3.12. Tailoring and Customization

4.1. System Design Processes
4.2. Stakeholder Expectations Definition
4.3. Technical Requirements Definition
4.4. Logical Decomposition
4.5. Design Solution Definition

5.1. Product Realization
5.2. Product Implementation
5.3. Product Integration
5.4. Product Verification
5.5. Product Validation
5.6. Product Transition

6.1. Crosscutting Technical Management
6.2. Technical Planning
6.3. Requirements Management
6.4. Interface Management

6.5. Technical Risk Management
6.6. Configuration Management
6.7. Technical Data Management
6.8. Technical Assessment
6.9. Decision Analysis

REFERENCES

1. Walden D., et. al., "INCOSE Systems Engineering Handbook", Fourth edition, John Wiley, 1994
2. Wasson, C., The Systems Engineering Conundrum – Where is the Engineering?, 17th Annual INCOSE international symposium, July 2021h
3. Sillitto, H., "Defining "System" – A Comprehensive Approach, 27th annual INCOSE Symposium,
4. Eisner, H. "Systems Engineering – Building Successful Systems", Morgan and Claypool, 2011
5. Senge, P., "The Fifth Discipline", Doubleday/Currency, 1990
6. Watson, M., and M. Griffin, "Systems Engineering: The Discipline of Engineering Elegance", see www.nasa.gov
7. Griffin, M., "How do we fix systems engineering?, 61st International Astronautical Congress, Prague, Czech Republic, Sept–Oct 2010
8. Eisner, H., "Systems Architecting", CRC Press, 2019
9. Seigers, N.J., et. al., "Learning from failure in systems engineering: A panel discussion", Wiley Periodical, 2011
10. Goldin, D., "Seeing the Angel", see Seaver, pepperdine.edu
11. MITRE, The Essence of MITRE'S systems engineering", systems engineering guide, 2010
12. NASA Systems Engineering Handbook, 2007

Operations Research: Revisited

6

This chapter deals with both the history and the future of operations research. It is generally accepted that operations research began in the UK and moved to the United States during WWI. Early applications had to do with military problems and involved the simplest of military issues, where the name was operational research. An earlier form of quantitative analysis as applied to warfare was the work done by Lanchester, known as Lanchester's equations. The formulation was in the form of differential equations of the form

$$dA/dt = -\text{Beta} \times B \text{ and } dB/dt = -\text{Alpha} \times A,$$

where A and B are the sizes of the A and B forces, respectively. We wind up with an understanding of what happens to force sizes A and B as a function of time, t. Although Lanchester's Equations are not necessarily associated with the early days of quantitative military analysis, it is true that operations research was indeed applied to military problems during the early days of Operations Research.

Lanchester's Equations have been applied to civil war battles such as Pickett's charge at Gettysburg, with good results. It has also been shown that groups of chimpanzees will not attack another group unless there is a numerical advantage of at least a factor of 1.5.

As the field developed, it was in the direction of more mathematics, as well as greater application domains. It is also fair to say that the following mathematical subjects can be viewed as having been part of the history of Operations Research:

    a. Mathematical Programming
    b. Network Analysis
    c. Dynamic Programming
    d. Game Theory
    e. Queueing Theory

DOI: 10.1201/9781003306610-6

      f. Inventory Theory
     g. Decision Theory
    h. Integer Programming
     i. Non-Linear Programming

Students of Operations Research will be expected to master the above theories as part of their MS or PhD curricula.

Organizations were formed with special homage to Operation Research. Examples are The Operation Research Office (ORO) and Operation Research Inc. which is the company that his author joined as the 22nd employee, back in 1959 (see chapter 9). The ORO, which was a non-profit, eventually became RAC, the Research Analysis Corporation.

The field of Operations Research was represented in the literature by a technical magazine with the not very inventive Operation Research title, which continued for many years. As time went on, there was a lot of pressure to somehow join the forces of Operation Research and Management Science with Operation Research, resulting in the organization with the name **INFORMS.**

The influence of radar on our anti-submarine warfare when we began to use Operations Research on military matters was strong. As an example, the German U-Boats had to revise their tactics and use of equipment, and that forced us to make changes in response to their changes. Operation Research continued to have an important role in anti-submarine warfare when looking at the security of our Fleet Ballistic Missile forces and their security. For example, the firm Operations Research Inc. had a contract to examine the potential vulnerability of these forces well into the 1980s. This firm "teamed" with the Applied Physics Lab at John Hopkins whereby the latter examined in detail the individual threats and ORI answered the questions – what does that all add up to in terms of changes in tactics and overall behavior? Thus, the focus was on operations – and how they might be adjusted to solve a particular problem.

Considerable attention was paid, in the early days, to issues of gunnery [1]. A problem of interest was posed as [Morse and Kimball] "suppose that a gun was fired at a target 100 times and that 40 hits were obtained. The appropriate question then becomes – what is the probability (p) that another shot will be a hit? The use of the binomial leads to the conclusion that the number 0.4 is a reasonable estimate of p. This is but one example of the application of probability theory to a military problem. In addition, a lot of attention was paid to defining measures of effectiveness, realizing that the wrong measures often led to the wrong answers. A classic questioning often led to the conclusion that we were solving the wrong problem. That possibility exists even today with the most sophisticated analysts.

Another area of analysis covered by Operations Research methods was that of estimating required force levels [m 7 k]. The following table appeared in this respect in the text from Morse and Kimball

MILES FROM BASE	CONVOYS	INDEPENDENTS	NAVAL VESSELS	TUFIL
0–100	4.3	17.4	2.0	1.4
100–200	2.1	**2.3**	0.9	**0.3**
200–300	0.6	1.7	0.7	0.1
300–400	**0.3**	**0.3**	**0.2**	0
400–500	0.1	**0.1**	0.1	0

These estimates of force level are interesting in themselves, but of special import is the number of independents required in the short range (up to 100 miles). Also of great interest is the analysis of Lanchester's Laws (both Linear and Square) under the overall subject of Strategical Kinematics {who knows what that might actually mean?).

Other military applications of Operations Research methods are as follows:

a. Problems of secrecy
b. Limitation of expert opinion
c. Selection of personnel
d. Search theory
e. Binomial theory
f. Poisson theory
g. Sampling theory

A random sampling of the Operation Research journal will reveal literally dozens of applications in the military arena (while avoiding the revelation of classified material). The same is true for the Naval Logistics Quarterly.

A BRIEF OVERVIEW OF INSTITUTIONS

If we go back to WWII, we see the distinct beginnings of Operations Research, mostly occurring in the UK in the domain of military analysis. Early work was carried out in regard to the use of Radar and increasing the time between radar's first warning and the attack by enemy aircraft. By 1942 formal operations research groups had been established in all of the UK's military services. The positive experiences in the UK soon caught fire in the

United States, in particular, the Naval Ordnance Lab in 1942. By the early 1950s, operations research had begun to be taken seriously in all military services, as well as industry. On the academic side, the first introduction of a formal degree in operations research took place in 1948 at MIT. Four years later, the Case Institute offered master's and doctoral programs in O.R. and the race was on. In the UK we see the following:

> The first scholarly journal (the Operation Research Quarterly (1950)
>
> The change of name to The Journal of the Operations Research Society of America (1952)
>
> The further change of name to Operations Research (1955)
>
> The initiation of the International Abstracts in Operations Research (1961)

We can see that a lot of solid commitment by way of studies occurred in the 1950s. Those were the days that serious foundations were laid in the field, leading to growth in both academic programs and industrial acceptance.

In moving from the Morse and Kimball text to the Hillier and Lieberman text we see a distinct change in coverage toward advanced mathematical structures. That is no great surprise in that such is the movement from the 1940s to the '60s and '70s. Topics identified by this author as advanced included:

a. Network analysis
b. Dynamic programming
c. Integer programming and
d. Non-linear programming

THE OPERATIONS RESEARCH OFFICE

An important milestone in the history of Operations Research was the establishment of the Army's Operation Research Office in 1952. This was a recognition of the value of Operations Research along with specific problem-solving to prove the point. Although the ORO was dis-established in 1961, it was essentially replaced with RAC (The Research Analysis Corporation).

If we move forward from the earliest days of OR we find the problematic topic of soft OR. In one incarnation, this is a small organization with very large clients, and therefore quite large impacts. The soft OR team consists of the following three people:

MIKE MCDONAGH (CEO AND CTO)

1. Susan (CFO and Chief Psychologist)
2. Steve Handy (Principal Agile consultant)

Mike is described as "some way between a cyber geek, agile geek, psychology geek, and cloudy architecture/programming geek". Mike and his team focus on making the business of software development safer, healthier, faster, cheaper, and happier. Mike is an AWS Certified Solution Architect. Susan combines skills as CFO and chief psychologist. She has a strong scientific background, working in public and private sector organizations. Finally, Steve Handy is a recognized expert in the field of agile software development. He has had many successes with agile software development, using tools to support systems and service development. A Soft OR team operates on the basis of three features:

a. Experience
b. Excellence
c. Empiricism

Another paper [2] deals with Soft OR and the contribution of the founders of Operations Research. The latter includes some 43 founders and considers the links between soft OR and these founders. The proposition is set forth that soft OR is a legitimate branch of Operations Research. Moving from this particular case of Soft OR, we find that there is a body of knowledge and approach that constitutes "Soft OR" (11), especially in a defense setting. Soft OR uses methods that are primarily qualitative, rational, interpretive, and structured so as to interpret, define and explore various aspects of certain classes of problems. Checkland's Soft Systems Methodology is considered to be an element of Soft OR [3].

CHECKLAND'S SOFT SYSTEMS METHODOLOGY [3]

The claim is that SSM can be very useful when, indeed, there are divergent views about the nature of the problem being addressed The methods of SSM have been most closely associated with systems thinking, including critical systems thinking.

There are some seven activities that have characterized SSM [4]. These are re-iterated below:

a. Entire situation considered to be problematical
b. Express the problem situation
c. Formulate root definitions of relevant systems of purposeful activity
d. Build conceptual models of the systems named in the root definitions
e. Compare models with real-world situations
f. Take action to improve the problem situation

Further research on SSM revealed that the pneumonic CATWOE gave us additional insight into what is meant by the SSM. The mneumonic CATWOE stands for the following:

- Customers
- Actors
- Transformation Process
- Worldview Owner
- Environmental

Looking more deeply into the obscure and unclear is, of course, a favorite topic of researchers. Such is the case with Checkland's SSM, which has some seven stages, as below:

1. Enter situation considered problematical
2. Express the problem situation
3. Formulate root definitions of relevant systems of purposeful activity
4. Build conceptual models of the systems named in the root definitions
5. Compare models with real-world situation
6. Define possible changes which are both possible and feasible
7. Take action to improve the problem situation

The answer itself is unclear, but there appear to be two logical answers to that question.

One answer is to suggest deeper investigations into CATWOE. The other is to consider returning to a place of comfort, which in this case is the mathematics of operations research. For the former approach, we have the topic of appreciative inquiry to consider.

APPRECIATIVE INQUIRY (AI) [4]

This relatively new field of investigation might be considered a part of operations research or organizational development. It is, in any case, about approaching change in organizations in a holistic and appreciative manner.

MOVING INTO 2022

As we move into the year 2022, we find the new INFORMS president, Radhika Kulkarni, delineating some top three new opportunities, as re-stated below [5]

1. Working together as one team
2. Following the all-encompassing strategic plan
3. Advocacy and promoting our profession

Ways to reach our goals within INFORMS were articulated as:

- Expanding and improving all core activities
- Partnering with other organizations
- Providing volunteering engagement opportunities
- Engaging and leveraging our subdivisions in a holistic manner
- Embracing operational motivation
- Creating an engaged and inspired staff environment

The scope of the strategic plan was defined as:

"the plan covers and embraces the full breadth and disciplines in the INFORMS community that are relevant to the vision and mission. As such, this includes operations research, analytics, management science, economics, behavioral science, statistics, artificial intelligence, data science, applied mathematics, and other relevant fields. We note the explicit inclusion of the two relatively new subjects of analytics and data science".

REFERENCES

1. Morse, P., and G. Kimball, "Methods of Operations Research.", MIT Technology Press and John Wiley, 1950
2. Heyer, R., "Understanding Soft Operations Research: The Methods, their application and its future in the defense setting", Information Systems Laboratory, DSTO-GD-0411, Australia, 20004
3. Checkland, "Soft Systems Methodology",
4. Watkins, J., and B. Mohr, "Appreciative Inquiry", Jossey Bass, 2001
5. Top ten opportunities from new INFORMS president, Radhika Kulkarni, INFORMS Today, February 2022

Common Elements – Operations Research and Systems Engineering

7

With the large scope of both operations research and systems engineering we would expect considerable overlap in the form of common elements. That is indeed the case, with a short form citation of such elements as:

- Modeling and Simulation (M & S)
- Optimization theories
- Software for the above
- Probability and Statistics applications
- Reliability theory
- Decision Theory
- Cost-Effectiveness Analysis
- Search Theory
- Forecasting
- Management

DOI: 10.1201/9781003306610-7

MODELING AND SIMULATION

This important topic allows the user to mirror the behavior of a system and deal with the model vs the actual real system, which may not yet exist. There are basically two types of models – the time step model and the event-based model. The choice as to which one is appropriate has to do with the nature of the problem that is being investigated. The event-based model is asynchronous and events determine the sequence of activities of interest. For example, in a simulation in the background of this author, the passing of a satellite over a set of ground stations was simulated. Events occurred in time, but not synchronously. As the satellite passed over each ground station, the time was recorded and the satellite went through a routine of activity with respect to that ground station. When the satellites were part of a constellation of data relay satellites, ground stations could be replaced by satellites themselves.

This is a deep and varied topic whereby the user models, or simulates, the behavior of systems (i.e., operations of a system) with the objective to see how to improve system behavior. Considerable software is available, off-the-shelf. Examples of such software is provided in the list below (Table 7.1):

TABLE 7.1   Selected Software Available for Modeling and Simulation (M & S) [1]

a. SysML
b. GPSS
c. Dynamo
d. Solid Edge
e. SimScale
f. FlexSim
g. SIMUL8
h. Arena
i. GASP
j. SLA

Some of the software that advertises the application of optimization routines includes the following (Table 7.2):

TABLE 7.2 Selected Optimization Software (Free and Open Source)

a. ADMB – NON LINEAR OPTIMIZATION USING AUTOMATIC DIFFERENTIATION
b. ASCEND – MATH MODELING; CHEMICAL PROCESS MODELING SYSTEM
c. CUTEr – TESTING ENVIRONMENT FOR OPTIMIZATION AND LINEAR ALGEBRA SOLVERS
d. GNU Octave – HIGH LEVEL PROGRAMMING LANGUAGE
e. Octeract Engine Community – PARALLEL DETERMINISTIC GLOBAL MINLP SOLVER
f. Scilab – CROSS PLATFORM NUMERICAL COMPUTATION PACKAGE

The reader is invited to do a google search of the above software to find out more in respect to the problem at hand.

PROBABILITY AND STATISTICS APPLICATIONS

There are a very large number of areas in which probability and statistics may be applied, to include design of experiments, making inferences and estimating relationships. A more productive approach is to look at a narrower field, namely reliability, maintainability, and availability (RMA). Important relationships for the first and third of those cited above are below, followed by several probability relationships [2].

SERIAL RELIABILITY

$R(\text{serial}) = \text{Reliability} = exp(-\text{\textsterling}t)$

$R(\text{total}) = R(1)R(2)\ldots\ldots R(n)$

PARALLEL RELIABILITY

$R(\text{total}) = 1 - [1 - R(1)][1 - R(2)]$

AVAILABILITY – GENERAL

The availability of a system can be calculated as:

A = MTBF/(MTBF + MTTR)

Where MTBF = mean time between failure and MTTR is the mean time to repair

Covariance; $COV(xy) = E(xy) - E(x)E(y)$

Correlation Coefficient = $cov(xy)/\text{sigma}(x)\text{sigma}(y)$

Mean Value of Sum = $E(x + Y) = E(x) + E(Y)$

AND Probability = $P(AB) = P(A|B)P(B) = P(B|A)P(A)$

Inclusive or probability = $P(E + F) = P(E) + P(F) - P(EF)$

Exponential distribution: $(\exp)^{-\pounds t}$

BINOMIAL DISTRIBUTION

$$(x + a)^n = \sum_{k=0}^{n} \binom{n}{k} x^k a^{n-k}$$

REDUNDANCY

We employ redundancy in our systems when we wish to avoid single-point failures. An example of the improvement we obtain from redundancy can be shown with Table 7.3 as follows:

TABLE 7.3 Effects of Redundancy on Reliability (Parallel Configuration)

RELIABILITY (R)	$R = 1 - (1 - R)(1 - R)$
.95	.9975
.9	.96
.8	.96
.7	.91
.6	.84
.5	.75
4	.64
.3	.51

So if we start with basic reliability of .95, that can be increased to .9975 by using simple redundancy. Similarly with basic reliability of .9 and .96.

DECISION THEORY

According to Holloway [3], the goal of his treatise on Decision Making Under Uncertainty is to present methods, concepts, and ideas of decision analysts at a level that can be understood by students, managers, and analysts who do not have extensive backgrounds in mathematics (algebra is sufficient).

In its most primitive form, decision theory may be viewed as a procedure for selecting among a set of well-defined alternatives. In that context, cost-effectiveness is a perfectly good way to do decision analysis. So one looks at the alternatives and characterizes each by its cost and its measure of effectiveness, and goes on from there.

Decision theory may be applied to problems in systems engineering and also operations research. For the former, for example, we may be using decision theory to decide upon which of several architectures is the correct one for a particular scenario. In that situation, decision theory may be applied in a cost-effectiveness context. In the field of operations research the example might come from a manager trying to pick the best route for deliveries of mail orders.

It should be noted that decision theory is one of the most basic elements of operations research, i.e., the original definition of operations research called for the type of analysis that would assist the decision maker in making decisions about his or her business operations.

Another approach can be found in the so-called Blackwell Handbook of Judgment and Decision Making [4]. Another descriptive phrase is the making decisions under uncertainty in which the latter is quantified rather precisely.

ANALYSIS OF ALTERNATIVES

The DoD has decided provided to formalize the matter of the analysis of alternatives (AoA). This is a special form of decision analysis, one in there are two or more alternatives one is considering to build or support.

COST-EFFECTIVENESS ANALYSIS [2]

Cost-effectiveness analysis can be considered a special form of decision analysis. Under this construct, we define very carefully at least two alternative decisions. They can be purchases or system developments or the like. Then the costs of each alternative are calculated along with measures of the effectiveness of the alternatives. In the simplest case, if the costs are lower and the MOEs are better, we select that alternative. If the costs are higher and the effectiveness superior, it may be that one is willing to pay the extra amount in order to obtain a higher level of effectiveness – or not.

As an example, for military systems, it is almost always the case that we are willing to spend the extra 21 dollars in order to achieve battlefield superiority, under the principle that we never wish to go into battle at a disadvantage.

A formal structure for cost-effectiveness analysis is the matrix wherein we rate each alternative against a set of criteria, each of which is weighted. We then calculate the ratings times the weights and sum the results. This sum represents a weighted effectiveness measure, which is then looked at against the costs of the various alternatives. This simple "rating and weighting" scheme is usually sufficient for a major portion of problems of his general type. This method is just about the same as what the military has called an AoA [5].

LIKELIHOOD RATIO

The computation of the likelihood ratio is a specific way to make a decision in the domain of digital signal detection. The overall theory calls for the calculation of the likelihood ratio as follows:

Decide on D_j if $p(y_j/x_1)/p(y_j/x_2) > p(x_2) (C_{21} - C_{22})$, decide on D_1, otherwise decide on D_2

$$P(x_1)(C_{12} - C_{11})$$

This example may be interpreted in terms of detecting a signal in noise where S is the signal and N is the noise power. Further considerations having to do with cost lead to a threshold boundary calculation of 9.39 volts [see ref. 1, p. 351].

SEARCH THEORY

Search theory shows up in operations research when, for example, we are searching for an optimum or dealing with a variety of coin-weighing problems, or trying to find an enemy submarine. In systems engineering, an example might have to do with searching for a cost-effective system solution or doing surveillance in an air defense scenario. We will borrow an example of search theory from one of the author's previous texts [1], as below. We start with the notion that we have tossed two keys into two tin cans, and we are at a point in which we are searching for the desired key. We assume that the desired key is twice as likely to be in one can (the red can). Thus the a priori probabilities are

P(red can) = 2/3 and P(blue can) = 1/3.

Assume further that the detection probability as a function of time is

$P(t) = 1 - exp(-£)t$.

For this situation, it can be shown that an optimal allocation for a total available time of, say, four hours is

- allocate a search time to the red can of 2 + (1/2) ln 2;
- allocate a search time to the blue can of 2 − (1/2) ln 2;
- we see that the total allocated time is four hours, and more time is spent looking in the red can which has a higher a priori probability.

LaGrange Multipliers: This is a specific way of searching for an optimum. We set up the problem and write down a set of equations, for example, involving the entropy of a system with n equal probabilities, as follows:

$$H + £[p(1) + p(2) + p(3) + ... P(n)\} − 1$$

Differentiation with respect to the $p(i)$ yields the set of equations

$$− [log \, p(1) + 1] + £ = 0$$
$$− \{log \, p(2) + 1] + £ = 0$$
$$− [log \, p(3) + 1] + £ = 0$$
$$........$$
$$− [log \, p(n) + 1] + £ = 0,$$

which implies that $log \, p(i) = f − 1$ for all $p(I)$ and therefore $p(i) = p(1) = p(2) = 1/n$ and the maximum entropy is $log \, n$, as stated earlier.

Other direct optimization techniques can be found in the literature under the topics of linear, quadratic, integer, backtrack, and dynamic programming. When the method is not clear or obvious, we at times resort to heuristic programming. And with this procedure, we select what we think is a local optimum in a step-by-step fashion. This is usually a good procedure, but not provably optimum.

FORECASTING

One might establish that there are five types of forecasting models, namely:

    a. Time Series Models
    b. Econometric Models
    c. Judgmental Forecasting Models
    d. The Delphi Method
    e. Artificial Intelligence (AI) Methods

Time Series Models: One may use a spreadsheet to great advantage in producing a time series. The algorithm is built in and one enters a series of data points (e.g., dow jones' values as a function of time in days. The spreadsheet will automatically generate new day values of the Dow Jones when clicking on Forecast Sheet.

Econometric Models: One such rather well-known econometric model is attributed to Leontief and relates the economics of the various sectors of our economy to all the other values, by sector.

Judgmental Forecasting: Used when there is little to no data available. As an example, economists may be asked to predict when the interest rates will get to a prescribed value.

The Delphi Method: This procedure uses a group of experts in a particular field. They are asked for their judgments in a series of rounds, whereby the results are shared and tend to converge after three to four rounds. Convergence will occur, but not necessarily to the correct answer.

AI Methods: These methods are based upon a particular algorithm, which, in turn, is based upon some number of real data points from the past.

The need for a forecasting model is largely determined by the type of system being considered. For example, in the field of transportation, a forecasting model is almost always needed. An example of the need in transportation might be described in terms of the demand for airport services when considering a new airport in a particular city. The size and shape and location of the airport will be determined, in large measure, by the forecasted demand.

The forecasting "model" for three types of forecasts pertaining to transportation are as cited below in Table 7.4:

a. Passenger demand for air travel
b. Freight demand forecast
c. A transportation modal split.

TABLE 7.4 Examples of Forecast Models

$$D = D_B(F/F_B \times I/I_B)^{-0.7}(T/T_B)^{-0.36}$$

Where D = demand, paasengers
F = fare
T = trip time
B = base value for each variable

F is the fare, I is disposable income, and T is trip time, and b is the base value in each case, the base year value is required and the current year is compared against it.

The freight forecast model takes the form:

$$N_{ij} = Kij \ p_i p_j / R_{ij},$$

where N is the number of tons of freight flowing from origin i to destination j
Population for region iD = the population distance from i to j
R = Population distance from region i to region j
K, a, b, c = parameters to be estimated, in order to use model
Software tools for forecasting [6]:

- BoostUp
  - Smart
  - OnPlan
  - revVana
  - Collective
  - Salesken
  - PipeDrive
  - INTUENDI
  - ForecatX
- Really Simple Systems CRM

MANAGEMENT

Management, especially project management, is a common element in both systems engineering and operations research. That is, both fields do project management when the occasion rises. The four elements of classical project management are as follows:

1. Planning
2. Organizing
3. Directing
4. Monitoring

Planning: The construction of an overall plan for the project, including a master schedule
Organizing: Developing a work breakdown structure of tasks
Directing: Assignment of people to the various tasks
Monitoring: Continuous checking of progress on tasks, and re-assignment, if necessary

REFERENCES

1. Eisner, H., "Computer-Aided Systems Engineering", Prentice-Hall, 1988, page 186
2. Eisner, H. "Cost-Effectiveness Analysis", CRC Press, 2012
3. Holloway, C., "Decision Making Under Uncertainty – Models and Choices", Prentice-Hall, 1979
4. Koehler, D., and N. Harvey, "Blackwell Handbook of Judgment and Decision Making:", John Wiley, 2008
5. "Analysis of Alternatives", DoD Instruction 5000.84, Office of Director, Cost Analysis and Program Evaluation, August 4, 2020
6. Software tools for forecasting; see Wikipedia

REFERENCES

Growth in Systems Engineering and Operations Research

8

In this chapter we consider growth patterns for the systems engineer and the operations analyst. That is, given that you have the training and skills as a systems engineer or analyst, what are the most likely paths of growth from that position? Or, put another way, where does the systems engineer likely to find himself or herself several years down the road through some process of growth? We will project a growth path for the systems engineer and check to see what actual people have taken these projected growth paths.

The primary skills of the systems engineer will be taken to be knowledge and mastery of the so-called systems approach. This has been articulated as deeply conversant with the following ten notions as applied to systems [1]:

1. Establish and follow a systematic and repeatable process
2. Assure interoperability and harmonious system operation
3. Consider alternatives at the various steps of design
4. Create a robust and slow-die system
5. Satisfy all agreed-upon user/customer requirements
6. Assure system sustainability
7. Use advanced technology, at appropriate levels of risk
8. Consider all stakeholders and their concerns about the system
9. Design and architect for system integration
10. Employ systems thinking

The last item on this list, systems thinking, conveys the essence of the systems approach and is considered to be the fifth discipline, in accordance with Senge's [2] view of how the world works. Creative holism is perhaps the best way to sum up what is meant by systems thinking. But other descriptors have been used, for example:

- Integrated
- Generalized
- System-wide
- Fusion
- Top-level

This also translates into a broad perspective when problem-solving so that no factors are neglected when thinking about what the areas of concern might be. A systems engineer with this broad view is more likely to be open to various growth pathways leading to other well-established fields of operation. The bottom line with respect to this thought process is that the systems engineer has the following five fields to evolve to:

a. local management
b. local leadership
c. chief problem solver
d. chief systems architect
e. chief enterprise architect

MANAGEMENT

The skill sets of the systems engineer prepare him or her for a position in local management. This means not only is he or she well prepared but also that he is quite likely to be offered and accept a management position when and if such an opportunity presents itself. The competence and perspective of the systems engineer increase the likelihood that he or she will be offered the position of project manager. This position is often the starting point for one to enter the general field of management. The logic of this argument proceeds as follows:

- General management is looking for an individual that "delivers" on a project, meaning success in the important attributes of schedule, cost, and performance. When such is the case, there is a tendency to keep up this faith in the individual in an ever-increasing set of responsibilities and also quite noticeable is the ability of such a person to build and manage a team.

These are just the early steps of evolving into a leadership role, where leadership is that often-not understood how one achieves this exalted state, but one tends to know it when one sees it. Indeed, the lead is someone who continues to have multiple successes on a variety of assignments. Leadership has been defined by several people, including this author, who points to the following attributes of a leader [3]:

    a. Practical visionary
    b. Inclusive communicator
    c. Positive doer
    d. Renewing facilitator
    e. Principled integrator

PROBLEM SOLVER

A noteworthy exposition on what management is looking for when hiring a new employee claims that "problem-solving" [4] is the number one attribute that management seeks. Management wants to be able to go to this valued employee with a problem and feel confident that the problem will be addressed and ultimately solved.

CHIEF ARCHITECT

The skill sets of the systems engineer make him or her ideally suited to serve as the chief architect of the systems that the company in question has set out to build. The systems engineer spends a lot of time architecting systems, and that pays dividends in that overall domain. And an important one it is! The right architecture is worth its weight in gold, and the wrong architecture leaves one continually playing catch-up and never seeming to get there.

CHIEF ENTERPRISE ARCHITECT

Architecting a system is different from architecting an enterprise. The former is localized to a particular system that the enterprise is building whereas the latter deals with the structure of the enterprise itself. Specific ideas with respect to the latter are in text [5]. This person literally knows everything there is to know about the enterprise, and is able to define the "best" architecture for the enterprise. There are several "models" for the structure of such an architecture:

    A. The McGovern et al. model
    B. The Eisner model
    C. The Zachman model

Very briefly, the McGovern model is described by McGovern and colleagues [5] as having five elements:

    1. Business goals
    2. Organizational matters
    3. The qualities of the organization
    4. The data representations and practices, and
    5. Agile modeling processes

The Eisner model consists of the following:

    1. The internal perspectives of vision and culture
    2. The people dimension of high-performance teams and rewards and accountability
    3. The systems dimension of reengineering and the learning organization
    4. The differentiation and leverage dimension, and
    5. The external dimension dealing with customers and competitors

The Zachman model, which is well-represented as the rows and columns of a two-way table, is as follows:

Rows

    a. who
    b. what
    c. when
    d. where

    e. why
    f. how

Columns

    a. contextual
    b. conceptual
    c. logical
    d. physical
    e. detailed

Whichever model is preferred by the chief enterprise architect, such a person has complete command over the enterprise and is quite successful in implementing an enterprise architecture.

To summarize, this author believes that the systems engineer is well-positioned to grow into one or more of the following roles in a corporate structure:

    a. Project manager
    b. Leader
    c. Problem solver
    d. Chief architect
    e. Chief enterprise architect

LOOKING AT PEOPLE TO "VERIFY"

We pause here to cite various individuals that start out as systems engineers and wind up in one or more of the aforementioned roles where the growth pattern is more or less the one suggested here:

- Norman Augustine
- Eberhardt Rechtin
- Irwin Jacobs
- Michael Griffin
- Seth Bonder

It may be argued that all five of the above persons have grown from the early days of systems engineering (in operations research for Seth Bonder) to something even more valuable with considerably more responsibility.

Norman Augustine began his engineering life with a degree in aeronautical engineering and relatively rapidly grew into the role of leader with a company by the name of Lockheed Martin. This was a rather spectacular achievement, with many successes along the pathways to leadership. Here are some facts about Norman Augustine, in case the reader does not know his background:

- Assistant Secretary of the Army
- President of Martin Marietta
- Supporter of the Boy Scouts (Eagle Scout, himself)
- Advisory Committee – U.S. Space Program
- Received 34 Honorary degrees
- Distinguished Civilian Service Award (five times)
- President, Lockheed Martin
- National Medal of Technology
- Books: Augustine's Laws, Augustine's Travels
- Fellow, American Academy of Arts and Science

Eberhardt Rechtin was a superb systems engineer working in the domain of communications engineering. He could do it all – design a space-based communication system Deep Space Network, and build and run a team of sophisticated engineers. What follows is a selected list of accomplishments for Dr. Rechtin:

- Books: Systems Architecting, The Art of Systems Architecting (w/ M.Maier)
- Developer: Way to Architect Systems
- Significant List of Heuristics for Building Systems
- PhD, Caltech
- "Father" of Deep Space Network
- NATO Advisory Group Aeronautical R & D
- Director of DARPA
- Assistant Secretary of Defense for Telecommunications
- President, Aerospace Corporation
- Professor, Systems Engineering, USC

Irwin Jacobs started out as a doctoral student and systems engineer at MIT and wound up running a company called Linkabit and later evolving into a company by the name of Qualcom which was quite significant in size. Here are some facts about Dr. Jacobs:

- Doctor of Science, MIT
- Professor, MIT and USC

- Book: Principles of Communications Engineering w(Wozencrcaft)
- UCSD School, named for him
- Co-founded Linkabit w/Andrew Viterbi
- Awarded "Entrepreneur of the Year" from Cornell
- Pioneered CDMA Systems
- IEEE Alexander Graham Bell Award
- Fellow of AAAS
- IMEC Lifetime Innovation Award

Michael Griffin held a team lead position at the APL of Johns Hopkins University and wound up as the administrator of NASA. In between, he held the following positions: president and CTO of Orbital Sciences, president and COO of In-Q-Tel, and Under Secretary of Defense for Research and Engineering. Following is a list of selected information about Dr. Griffin:

- PhD, University of Maryland, six other degrees
- NASA Administrator
- Head, Space Dept., APL, Johns Hopkins
- Professor, University of Alabama, Huntsville
- Under Secretary of Defense, Research and Engineering
- Deputy of Technology, SDI Program
- President-Elect, AIAA
- Key Study – Extending Human Presence in the Solar System
- Included in TIME 100, most influential people

We will use a similar point of view when examining the field of operations research. First, we start with a skill set different from that of systems engineers but with a very strong capability in quantitative methods. The operations analyst became well-known bringing these special skills to bear on a variety of quantitative applications. Typically, the super technical research analyst really enjoys playing with mathematical constructs and interpreting the results. Experiencing growth, then, leads the operations research analyst to the likes of (a) modeling and simulation (M & S) and (b) analytics. Transitions to management in a technical enterprise, as per the systems engineer are not likely with this type of person. However, this author has seen this type of person agree to become Department Chair in two cases, with excellent results. So growth into management can easily occur when one is considering academia.

Next, we cite specific persons that have grown in the case of operations research individuals. The first is Seth Bonder, whose capability in operations research is well known. Seth grew from an individual analyst to forming and managing his own company. Typically, operations research analysts have not been known for their strength as entrepreneurs, but occasionally we notice that one has taken this unusual path; additional information about Seth Bonder is listed below:

- Earned PhD in Operations Research at Ohio State
- Founded Vector Research Inc in 1972
- Member, National Academy of Engineering
- Fellow, Institute for Operations Research and Management Sciences
- Awarded Vance Wanner Memorial Award
- Professor at the University of Michigan

The two people that grew from OR analyst to Department chairs have been Tom Mazzuchi, and Zoe Szainfarber, both from the EMSE Department at George Washington University. An entrepreneur that moved from a well-known analyst directly into a self-started firm was Emory Cook, who started Operations Research Inc., the firm that this author joined (see Chapter 9) back in 1959.

Another analyst that moved into an executive position, running a company was Ellis Johnson, who took on the presidency of the Operations Research Office (ORO). He ran that enterprise successfully for about seven years, on behalf of the Army. Johnson was known for integer programming and combinatorial optimization and received his PhD in Operations Research at the University of California at Berkeley. He taught at Georgia Tech and received the John von Neumann Theory Prize. He was a Fellow of three well-known organizations.

Yet another outstanding figure in operations research was George Dantzig who was known for his development of the simplex algorithm and generalized linear programming.

John D. C. Little is a well-known entrepreneur with a deep background in Operations Research. He taught at the Sloan School of MIT and is considered the founder of a field known as marketing science. He served as president of both ORSA and TIMS, and advanced the state-of-the-art in marketing research and analysis.

GROWTH OF THE ENTERPRISE

A usual pattern of growth occurs when, as the individual grows, so does the enterprise itself. This is clearly evident in the case of Dr. Jack Welch who served as president of GE from 1981 to 2001. During that period of time, GE increased its market value from 14 billion to 410 billion. He gained a reputation as neutron Jack, whereby he closed many profit centers or sold them to a third party. He passed away in 2020, at age 84.

GROWTH OF THE FIELD OF OPERATIONS RESEARCH

According to the Bureau of Labor Statistics [6], the overall field of operations research has been on a growth pattern since its inception. The median pay has gone up to $86,200, representing steady growth, with a breakdown by sector shown below:

SECTOR MEDIAN	ANNUAL RATE ($)
Federal Government	119,720
Manufacturing	94,340
Management of companies and Enterprises	91,000
Finance and Insurance	86,280
Professional & Scientific	85,950

- The median pay for operations analysts in 2020 was $88,200
- The total number of jobs in 2020 has been estimated as 104,100
- The total number of operations analysts is expected to grow 25% from 2020 to 2030.

GROWTH OF SYSTEMS ENGINEERS

- There are some 305,909 systems engineers currently employed in the United States
- The average age of a working systems engineer is 39 years
- Systems Engineers are most in demand in Chantilly, VA
- 79% of systems engineers are male, 16 percent female
- The average annual salary of systems engineers is $90,107
- The male income for systems engineers is $88,134
- The female income for systems engineers is $81,889
- The starting salary for systems engineers is $68,000

COMPARISONS WITH SIMILAR OCCUPATIONS

OCCUPATION	LEVEL OF EDUCATION	2020 MEDIAN PAY
Economist	Master's	$108,350
Industrial Engineer	Bachelor's	$88,950
Logisticians	Bachelor's	$76,270
Management Analysts	Bachelor's	$87,660
Market Research Analysts	Bachelor's	$65,810
Mathematicians and Statisticians	Master's	$93,290
Software Developers	Bachelor's	$110,140

REFERENCES

1. Eisner, H., The Systems Approach, p. 8, "Systems Architecting", CRC Press, 2020
2. Senge, P., "The Fifth Discipline", Doubleday/Currency, 1990
3. Eisner, H., "Reengineering Yourself and your Company", Artech House, 1990
4. Eisner, H., "Problem-Solving", CRC Press, 2020
5. Enterprise Architecting, CRC Architecting
6. U.S. Bureau of Labor Statistics, salaries for various occupations

Key Contributions – Operations Research

9

As suggested in other chapters in this treatise, the field of Operations Research has made seminal contributions of no small significance. Mainly, they have been in the arena of problem-solving as related to mathematical structures. We can take a top-level reading of some of these contributions by noting the subject areas in two of the early books on operations research [1,2], as follows:

BASIC CONTRIBUTIONS IN OPERATIONS RESEARCH [1]

LINEAR PROGRAMMING – SIMPLEX METHOD (GEO. DANTZIG)	SELECTED TOPICS IN OPERATIONS RESEARCH [2]
Special Algorithms for LP	Measures of Effectiveness
Duality Theory	Strategical Kinematics
Goal Programming	Tactical Analysis
The Assignment Problem	Gunnery Problems
The Transportation Problem	Bombardment Problems
Network Analysis	Operational Experiments
Dynamic Programming	Organizational Problems

(Continued)

DOI: 10.1201/9781003306610-9

LINEAR PROGRAMMING – SIMPLEX METHOD (GEO. DANTZIG)	SELECTED TOPICS IN OPERATIONS RESEARCH [2]
Game Theory	
Queuing Theory	
Inventory Theory	
Decision Analysis	
Integer Programming	
Non-linear Programming	

IMPACTS IN TECHNIQUE AREAS [3, 4]	NO OF PROJECTS	FREQUENCY OF USE
Statistical Analysis	63	29
Simulation	54	23
Linear Programming	41	19
Inventory Theory	13	6
PERT/CPM	13	6
Dynamic Programming	9	4
Non-linear Programming	7	3
Queueing Theory	2	1
Heuristic Programming	2	1
Miscellaneous	13	6

RELATIVE USE OF OPERATIONS RESEARCH METHODS	RESPONDENTS(*)
Regression Analysis	74
Linear Programming	78
Simulation	70
Network Models	69
Queuing Theory	71
Dynamic Programming	69
Game Theory	67

(*) Ledbetter/Cox Survey.

ARMY'S OPERATIONS RESEARCH OFFICE

A key organization devoted to matters of Operations Research was the army's Operations Research Office (ORO) established in 1952 in Chevy Chase, Maryland. That office was disestablished in 1961 and was effectively replaced by RAC, the Research Analysis Corporation. That organization existed until 1960.

FEDERALLY FUNDED R & D CENTERS

There are several FFRD Centers that are focused on Operations Research. Two of them are:

- The Homeland security Operational Analysis Center, and
- The National Defense Research Institute, run by the RAND Corporation in Santa Monica

MILITARY MATTERS

It is well-known that Operations Research contributed mightily to the conduct of various forms of warfare, starting in the UK. Areas of application included:

- Antisubmarine warfare
- Radar systems
- Search theory (surveillance for targets in air defense scenarios)
- Principles of command and control in air defense scenarios
- Game theory (international negotiation)

KEY CONTRIBUTORS

There have been quite a few contributors to the field of operations research, including the list below [4], together with a very brief citation of some of their contributions:

CONTRIBUTOR	CONTRIBUTION
Arjang Assad	Professor at University of Buffalo; contributor of many papers
Russ Ackoff	Pioneer in overall field
E, Leonard Arnoff	Pioneer in overall field
Stafford Beer	Pioneer in Artificial Intelligence and Cybernetics
Patrick Blackett	Nobel Laureate in Physics; Fellow of Royal Society of Operational Research
Alfred Blumstein	Pioneer in studies of public policy
Seth Bonder	Started own company in Operations Research; carried out many applications, especially on military matters
Abraham Charnes	Mathematical applications to field
C. West Churchman	Important contributor to overall field over six decades; seminal book
William W. Cooper	Developed conspiracy Theory
George B. Danzig	First recipient of ORSA award; developed algorithm for simplex method
Jay Forrester	Formulated theory of industrial dynamics
Ray Fulkerson	Formulated optimization algorithms
Saul Gass	Contributor of many articles and overall field
Murray Geissler	Pioneer in study of flow of goods in organizations
Ralph Gomory	Successful pioneer in University setting
Ron Howard	Formulated theory of dynamic Programming And Markov Processes
Ellis Johnson	Served as Director of the Operations Research Office (ORO)
George Kimball	Together with Morse, wrote seminal book on Operations Research
Harry Markowitz	Known as driving force for computer simulation model/software
John D. C. Little	Successful leader in industry studies in Operation Research
Philip Morse	Founding Member and first president of ORSA; wrote seminal book with G. Kimball
Thomas Saaty	Wrote seminal book; formulated AHP Process Model
Herbert Simon	Wrote seminal book on AI; important mathematical economist
John von Neumann	Significant contributor to Artificial Intelligence
Steven Vajda	Brought linear programming to US and UK
Evelyn Martin Beale	Significant contributor to mathematical theory

CONTRIBUTOR	CONTRIBUTION
James Matheson	Developer of Markov decision processes
Jacinto Steinhardt	Solved important military problems
Hugh Miser	Solved important military problems
George Kozmetsky	Visionary in Operations Research and Management Science

SOFT OR AND PRACTICE: CONTRIBUTIONS OF FOUNDERS

A particularly cogent overview of contributions of the founders of OR addresses the matter of "soft OR" [5]. This involves such matters as belief based on faith rather than evidence; impacts on strategic social issues; and intangible and not easy-to-explain approaches. Soft OR and Checkland's [5] ideas have often been connected to one another.

REFERENCES

1. Morse, P., and G. Kimball, "Methods of Operations Research", MIT Technology Press and John Wiley, 1950
2. Hillier, F., and G. Lieberman, "Introduction to Operations Research", Third Edition, Holden_Day, Inc., 1980
3. West Churchman, C., R. Ackoff, and E. L. Arnoff, "Introduction to Operations Research", John Wiley, 1957
4. Beasley, J., "OR Notes". See http://people.brunel.ac.uk
5. Checkland, P. "Soft Systems Methodology", John Wiley, 2007

Key Contributions – Systems Engineering **10**

Systems engineering is a quite popular program of study, especially at the master's level. There are many job offerings that pertain to this area of study and work. So it is the perceived value of systems engineering that has gotten the field to where it has so much demand. The Software Engineering Institute has recognized this value by means of an article with the title "The Value of Systems engineering" [1]. The "value" of this field may be represented by the fact that systems engineering has made, over the years, important contributions. The INCOSE handbook [2] articulates some of the milestones in terms of contributions, as follows:

1937 – British Multidisciplinary team analyzes the air defense system
1939–1945 – Bell Labs supported NIKE missile project development
1951–1980 – SAGE air defense system defined and managed by MIT
1954 – Recommendation by the RAND Corporation to adopt the term systems engineering
1956 – Invention of Systems Analysis by the RAND corporation
1962 – Publication of "A Methodology for Systems Engineering" by A. D. Hall
1969 – Modeling of Urban Systems at MIT by Jay Forrester
1990 – National Council on Systems Engineering Established
1995 – National Council changed to INCOSE to broaden NCOSE scope and membership
2008 – ISO, IEC, IEEE, INCOSE, and others fully harmonize SE concepts
Important dates related to standards:
1069 – Mil-Std 499
1979 – Army Field Manual 770–78

1994 – Perry Memorandum urges military contractors to adopt military standards
1998 – EIA 632 released
IEEE 1220 released
1999–2002 – ISO/IEC 15288 RELEASED and adopted by IEEE in 2007
2002 – Guide to the Systems Engineering Body of Knowledge Released

SIGNIFICANT SE STANDARDS AND GUIDES

ISO/IEC/IEEE/15288 – Systems and Software Engineering – System Life Cycle Processes SEBoK – Systems Engineering Body of Knowledge.

The Director of Systems Engineering in the Office of Systems Engineering within the Office of The Secretary of Defense (OSD) has commented upon the value of systems engineering in relation to improvements in time, cost, and/or performance that could be achieved through the appropriate use of systems engineering notions and concepts. Similarly, in a report from the GAO [3] (government accountability office). However, the ideal is often not achieved, as noted in the paper by Michael Griffin – how to fix systems engineering [4]. Griffin (and his co-author) claim that what we are in search of is design elegance, which is a function of the designer and not the so-called process. The words used by these authors are "a broader view of the systems engineering discipline is required, one that encompasses the notion of elegance in the design as a specific goal of systems engineering". The four features of an elegant design, according to these authors, are as follows:

1. Efficacy
2. Robustness
3. Efficiency
4. Unintended consequences

Design elegance appears to be a goal and it's not obvious as to how to achieve this form of nirvana. Or, in other words, systems engineering can indeed be used to improve the development of large complex systems, but only after we have mastered the matter of how to achieve design elegance. A similar view regarding systems engineering was expressed by Robert Frosch when he was administrator of NASA [5]. That view can be summed up with the following observations:

a. The so-called tools of systems engineering may well be part of the problem rather than the solution
b. These tools include milestone charts, PERT diagrams, configuration management, and the like
c. Massive amounts of paper are often generated in response to problems with systems, such that the generation of all that paper actually interferes with solving the problem(s)
d. Linearity is assumed rather than the real-world issues of the system in question
e. The statement of the true problem is often wrong and needs to be examined in greater detail to get to the real root cause of the problem
f. We may well have lost sight of the fact that engineering is more of an art, rather than a science, as advertised

The long and short of this issue is that we need to call in the truly imaginative manager who has a feel for the reality of the problem and its solution and trust the highly intuitive approach to problem-solving.

So – we need to accept the notion that many of the tools of systems engineering are to be replaced with true engineering vs. using so-called management tools. This is a difficult notion to accept that fixing systems engineering approaches will be fraught with lack of assumptions and uncertainties.

So the answer lies in finding the correct problem solver and not in finding the right systems engineering management tool or tools. So much for "fixing" today's version of systems engineering.

CONTRIBUTIONS TO MISSILE SYSTEM DEVELOPMENTS

If we look back at the history of some of the companies on the West Coast, we see companies like TRW that had special influences on the development of missile systems and technologies. That work, supported by the Air Force, set the stage for today's capabilities in national missile defense which, in turn, put this country in an advantageous position with respect to air defense missions. At the same time, East Coast activities in systems engineering helped to assure that submarine survivability was achievable in relation to a key element of our triad. So the bottom line is that Griffin and his co-author have given us some guidance in producing an elegant design. We re-state these as steps:

Step One – Functionally decompose the system (from this author)
Step Two – Consider the features of an elegant design, one at a time (see above)
Step Thee – From this author – Consider needed redundancies (for single point failures)

SYSTEMS ENGINEERING RESEARCH CENTERS

Significant contributions are made in Systems Engineering by the Systems Engineering Research Centers (SERCs), especially in terms of research problem areas, a sample of which is listed below for recent years (Table 10.1).

TABLE 10.1 Representative Sample of Research Problem Areas in Systems Engineering

- Systems of Systems
- Mission Engineering
- PEO Missiles and Space Engineering Methodology Implementation
- Comprehensive Enterprise/System of Systems (SoS) Modeling and Analysis
- System of Systems Analytic Workbench
- Approaches to Achieve Modularity Benefits in the Acquisition Ecosystem
- New Observing Strategies Testbed
- Digital Enterprise Transformation

The history of these SERCs goes back to March 2010 when John Baras led the winning team for a SERC at the University of Maryland. In more recent times, Dinesh Verma led a team of some 22 universities to establish a SERC at the Stevens Institute of Technology in Hoboken, New Jersey. This SERC is advertised as:

"this SERC, which is a University-Affiliated Research Center at the Department of Defense, leverages the research and expertise of senior lead researchers from 22 collaborator universities throughout the united states. The SERC is unprecedented in the depth and breadth of its reach, leadership and citizenship in systems engineering through its conduct of vitally important research and the education of future systems engineering leaders".

HEURISTICS OF SYSTEMS ENGINEERING

In his book on Systems Architecting, Eberhardt Rechtin set forth a serious number of heuristics that the systems engineer should take into account as he or she is building a system. The list below is a short explication of some of these heuristics:

1. A model is not reality
2. Keep it simple, stupid (KISS)
3. The simplest solution is usually the correct one.
4. In partitioning a system into subsystems, choose a configuration with minimal communication between the subsystems
5. Sometimes, but not always, the best way to solve a difficult problem is to expand it
6. Extreme requirements should remain under challenge throughout system design, implementation, and operation
7. Work forwards and backwards
8. The choice between architectures may well depend upon which set of drawbacks the client can handle best
9. No complex system can be optimized for all parties concerned, nor all functions optimized
10. Mid a wash of paper, a small number of documents become pivots around which every project's management revolves
11. If it ain't broke, don't fix it
12. Quality cannot be tested in, it has to be built in

GETTING A MAN TO THE MOON AND OTHER EXPLORATIONS

We have to understand how big an undertaking it was getting a man to the moon, and that it could not have been done without the discipline of systems engineering. The same is true for our various space explorations as well as inventories of the earth from space, otherwise identified as earth resources technologies using instruments such as tv vidicons and spectrometers.

CONTRIBUTIONS TO INFORMATION SYSTEMS ENGINEERING

A subset of topics in which we will be making progress in the next couple of decades is that of information systems engineering [6]. A Springer book focused on this body of knowledge, highlighting seminal papers in the so-called CAiSE arena. Ten noteworthy papers are cited below in Table 10.2.

TABLE 10.2   Ten Selected Papers in The Field of Information Systems Engineering

- The CAiSE adventure
- A natural language approach for requirements engineering
- Conceptual modeling and natural language analysis
- The three dimensions of requirements engineering: 20 years later
- 20 years of quality of models
- MetaEdit+ at the age of 20
- Workflow time management revisited
- Promises and failures of research in dynamic service composition
- On structured workflow modeling
- Evolution of the CAiSE author community: a social network analysis

MODEL-BASED SYSTEMS ENGINEERING

Another contribution to the field has been that of model-based systems engineering. One might say that Prof. Wayne Wymore set forth this notion which taken hold and at this time represents a well-developed aspect of systems engineering. The apparent leader in this regard is Sandy Friedenthal who has authored the significant text cited here [7].

METHODOLOGY FOR VERY LARGE-SCALE PROGRAMS

This contribution has given us the confidence and wherewith all to do all that it takes to send man to the moon. Another example is that of National Missile Defense, which started out as the SDI (Strategic Defense Initiative Initiative).

This contribution is related to the theory of Systems of Systems. Although there have been problems with the overall methodology from time to time, the systems engineering approach, as of today, has allowed us to deal with seemingly impossible scenarios, especially with respect to national missile defense. In this domain, we are talking about "hitting a missile with a missile", we have a great need to make this work, and our systems engineering approach contributes in a very important way to the likelihood of success.

KEY CONTRIBUTORS

Below is a list of key contributors to the field of systems engineering:

CONTRIBUTOR	CONTRIBUTIONS
Sarah Sheard	Active researcher in the field of complexity analysis and management
Eberhardt Rechtin	Wrote a seminal book on Systems Architecting; Lead professor and President pf Aerospace Corp
Mark Maier	Wrote seminal text with Rechtin; senior consultant
Sandy Friedenthal	A leading researcher in MBSE
Simon Ramo	Ran large company (TRW) and served as a consultant to businesses
Admiral Rickover	Developed Navy's nuclear Navy, including missile systems
Robert Oppenheimer	Managed Manhattan Project for Atomic Bomb
Andrew Sage	A major contributor to literature; wrote several seminal books
Ben Blanchard and Wolt Fabrycky	Deceased, worked with Wolt Fabrycky Professors at V. Tech; wrote seminal text w/Ben Blanchard
Henry Petroski	Prolific Writer of wisdom in engineering
Norman Augustine	Prior CEO of Martin-Marietta; consultant to business and government
Hewlett and Packard	Built the leading edge firm (technology, management) of HP
Andy Grove	Prior Director and Executive at INTEL Corp.

(Continued)

CONTRIBUTOR	CONTRIBUTIONS
Stephen Jobs	Genius of Product Design; Very successful company developer
Benjamin Franklin	Old School Inventor and ambassador
Thomas Edison	Genius at invention; large number of patents
Kelly Johnson	Managed "skunk works' at Lockheed; built successful aircraft
Tim Brown	Formulated theory of design
Alexander Kossiakoff	Technical Director of JPL labs, contributed to national defense programs
Mervin Kelly	Managed Bell Labs
Michael Griffin	Prior Administrator, NASA; professor
J. Edwards Deming	The genius behind TQM (Total Quality Management) Concept

INTERNATIONAL COUNCIL OF SYSTEMS ENGINEERING

The systems engineering community has built the International Council of Systems Engineering (INCOSE) which now has some 19,400 participating members. INCOSE has been forward-looking and very strong in membership. INCOSE has provided a vision for the future [8], which simultaneously looked both forward and backward with the following contributions:

a. Articulating The changing nature of future systems, oriented to solving a "diverse spectrum of societal needs"
b. The above, in part, will require the transition to a model-based discipline
c. Transformations in SE to meet future needs

INCOSE has recognized, in this vision document, the fact that problems of the future are likely to be within the "diverse spectrum of societal needs". Solving some of these problems (i.e., global warming and its effects) can make the difference between survival on this planet or not. Placing such a burden on INCOSE is not reasonable, but we can expect an increasingly important role on the plate of INCOSE, as we tackle literally survival issues for the future.

If we follow the logic of the vision report, we see special attention being paid to "the future state of systems engineering" (part 3 of the document). This section delineates possible future contributions to include the following:

a. Impacts of the digital transformation
b. Impacts of AI
c. Systems engineering practices
d. SE theoretical foundations
e. Expanded applications

The bottom line in this regard is simply that a vision for the future of systems engineering is simply to address future problems and make significant contributions when doing so.

The INCOSE handbook [8] itself represents a contribution to the field. It is comprehensive and quite well-written and serves as an extremely good source.

REFERENCES

1. "The Value of Systems Engineering", SEI Blog, Software Engineering Institute, May 20, 2013
2. Walden, D., et. al. "INCOSE Systems Engineering Handbook", Fourth Edition, INCOSE, John Wiley, 2016
3. GAO report, cost estimating and assessment guide, 2020
4. Griffin M., and L. Watson-Morgan, 8 March 2020, Alabama Hall of Fame Introduction
5. Frosch, R., "A Classic Look at Systems Engineering", IEEE international convention on Aerospace and Electronic Systems, New York, March 26, 1069
6. CAiSE, see www.nsta.org
7. Friedenthal, S., "SysML: The Modeling Language, a Practical Guide to SysML"
8. INCOSE systems engineering Vision 2035, "A World in Motion", INCOSE, 2020

A Personal Journey

11

This author has spent the better part of 60 years in the fields of operations research and systems engineering. Trained as an electrical engineer (three degrees), the latter field has received the most attention, but not without considerable time in the former. This chapter discusses adventures over the years in both fields.

RECOGNIZING OPERATIONS RESEARCH

There came to be a time in this author's life when, literally, it was time to get a job to support a new family. It was the year 1958 and a new child was scheduled to appear in early 1959. So I started looking seriously for a job in the industry, recognizing that the salary of a lecturer in physics was not sufficient. I had been teaching physics at Brooklyn College and picked up a copy of the New York Times where I found an ad looking for a person to join the staff of Operations Research Inc., located in Silver Spring, Maryland. I answered the ad and found myself being interviewed for a research position at the company a few weeks later. The interview went well despite the fact that I had no background in operations research. Instead, I had a master's degree in electrical engineering just that year from Columbia University.

I soon found myself driving off to Silver Spring, Maryland with my wife (in the front seat) and baby (in the back seat) to introduce this early family to a new city. With whatever spare time I had, I was looking into the field of operations research, since that was the name of my new company. I distinctly remember reading (actually skimming) the Kimball and Morse book on this field, and wondering – is this what it's all about? And if so, it was clearly new to me, but of considerable interest. So I soon found myself being introduced to my new boss, who gave me my first assignment: to determine the performance of our torpedoes in shallow water. I began to read all about the subject – about torpedoes, about how they worked in deep water, and about what was different about shallow water. I proceeded to construct what I called a mathematical model of the overall scenario, for two types of torpedoes. I eventually looked at the time variable gain of these torpedoes and focused in on the notion of "the locus of reverberation peaks" as the essence of the model. I carried out a fair amount of sample run analyses, and I remember pounding away at a desk calculator to get a variety of parametric answers. I presented all of my results to my boss, who seemed pleased with what I had done. The next step was to present these results to our Navy sponsor, located in the Navy building area of downtown Washington. That too went well, and so I was off and running in my new company, with a success story. Later that year, during review time, I got a hefty raise as well as praise for a job well done. Meanwhile, I was reading everything I could get my hands on about operations research.

My next assignment was more challenging, with mixed results. We had a small study contract with the Applied Physics Lab of Johns Hopkins University, to look into advanced radar detection techniques associated with a forthcoming air defense system. So the reading started again, this time about radars and statistical communications theory. There I was, still outside of operations research, but a whole lot closer to my original field of electrical engineering, at the grand old age of 24.

So in order to make more progress with this new assignment, I was moved into sharing an office with the project manager whose name was George. He was a very clever and pleasant PhD physicist whom I had a lot to learn from. I remember him as slightly eccentric since he used gold bricks to store and save his money.

But as it turned out, we had not been able to make much progress since the customer knew a whole lot more about the problem than we did. So we "came clean" and suggested termination of the project without prejudice. And that's exactly what happened, and I was left with no assignment for a short while. However, and fortunately, a project for the Federal Aviation Administration (FAA) came my way. It was for the system performance branch inside the FAA, with a very broad as well as unclear statement of work. This was par for the course in my job, and in many ways represented how problems were ill-defined in those days. Eventually, I was led to the question – how should we monitor and assure the performance of our Air Traffic Control (ATC) radar systems? My boss on that project (another PhD physicist) was not much help in defining how to address that issue. Eventually I developed an approach, using concepts from operations research as well as electrical engineering. The notion was that in order to assure performance over the long haul, we had to monitor performance on a daily basis. This meant making certain measurements every day and recording "out-of-bounds" values especially. So in order to prove my approach, I designed what I called a "radar quality control feasibility experiment", involving real-world measurements on a variety of radars in the field. Working with a new office mate, we went to the radar sites and made the measurements over a period, as I recall, of approximately 30 days. This was very satisfying since it involved getting out of the office and working with real ASR FAA radars, and real physical equipment. In any case, the project was very successful as our FAA sponsors concurred with our approach and tried to make some changes within the FAA, but without much success.

The president of my company had strong ties to The Applied Physics Lab at Hopkins and they were a lead laboratory when it came to advanced missile air defense systems such as the three T's (Terrier, Tartar, and Talos). So when the time came, I was asked to participate in a summer "study" involving the next version of a missile defense system. I had little background in this area but had to learn quickly in order to make a contribution. The tasking involved the generic "ASMS" title and was my first introduction to large-scale and large-capability systems. My later work on the Basic Point Defense Surface Missile System (BPDSMS) was assisted by these earlier days on the ASMS study.

The next challenge, as I recall, was to write a proposal to NASA/Goddard Space Flight Center, for a reliability assessment of the Nimbus meteorological satellite. I wrote most of that proposal, and we were awarded the contract. This was my first win, and also very satisfying both for the win itself as well as for the fact that it would involve a team of about five people. And so it became clear as to how the system worked – you not only had to do the work, you had to write proposals in order to obtain the work. Now that was above and beyond – but it was clearly a way to progress in a company, and in both operations research as well as systems engineering.

So we started out on the Nimbus reliability investigation, digging deeply into all the documentation for the various Nimbus systems – power supply, attitude control (3-axis stabilized), thermal control, communications, telemetry, and payload systems (e.g., Videcon, etc.). It was extremely challenging, partly because it was the first such technology for most of our team members, but also because we had to serve as consultants to Goddard engineers, and be knowledgeable in dealing with the prime contractor, who happened to be General Electric (GE) in Valley Forge.

We managed to survive that challenge rather well and finally documented our study in a massive report that "predicted" the numerical reliability of Nimbus but also highlighted reliability concerns in terms of the design of the sub-systems. So there was a real-world project that was more systems engineering than it was operations research – chalk up three years of work on satellite technology.

After our success with Nimbus I was made official project leader just about at the time when a new project came along. We became part of the team that bid on the Mallard battlefield communications system, for the Army folks at Fort Monmouth. The team consisted of G, T & E – Sylvania in Waltham, MA, IBM, and us. We won the contract and all of a sudden I had two simultaneous jobs – project leader on Nimbus and also project leader on Mallard. Apparently I had shown enough scope and work dedication to believe I could handle two project leadership assignments simultaneously. I was progressing in both fields but still felt that I needed some more technical training. So I applied for two new "jobs" – teaching double "e" at George Washington University, and also enrolling in their doctoral program that involved five fields of study (it was the engineering school that had no operations research department at that time). The five fields were (1) probability theory, (2) communications theory, (3) information theory, (4) non-linear systems, and (5) statistics. So that made four jobs – two at work (Nimbus and Mallard) and two at GW – adjunct assistant professor and student. It was, as I recall, a difficult but very productive time.

The work at GW paid immediate dividends in terms of an ever-expanding scope of work at Goddard. That included giving courses in advanced topics (communication engineering) that related directly to my five fields of study. And that evolved into yet another task as a project leader – work on the IRLS (interrogation, recording, and location system) at Goddard and after that, project leadership on the AOSO (advanced orbiting solar observatory) system. That work was much the same as the Nimbus scope, except there were new and different payloads. After about a year, the AOSO project was canceled at Goddard, which was a relief since I was distinctly in overload, especially at Goddard.

Throughout all of the above work with the FAA and with Goddard and Mallard, I was learning how to be a project leader. This was quite demanding but I seemed to accept the challenge. But it was time away from technical content, in the main, and was not my forte or preference at the time.

I had kept in touch with the FAA, and when the time came to write a proposal to build a "NAS Model", I jumped at it. With a very strong technical approach and acceptable cost, I wrote a winning proposal and was soon off and running building such a model. The approach, in the main, involved structuring a "model of models", and we spent the better part of two years working on that. It was very successful when the internal machinations at the FAA led to the cancellation of the remaining work tasks. But this brought me back into serious modeling activity which was, no doubt, part of both systems engineering as well as operations research. The NAS Model was extensive, and consisted of the interplay and interaction of some 12 sub-models (excluding cost), namely:

1. Airport capacity
2. Airspace capacity

3. Delay
4. ATC Availability
5. Trip Time
6. Energy Utilization
7. Service Availability
8. Noise
9. Air Pollution
10. Security
11. Safety
12. Demand

Meanwhile, my work on Mallard allowed me to very closely observe what G, T & E did as systems engineers on Mallard. It indeed formed the basis for an independent "study" of the essence of systems engineering. That activity impressed upon me the importance of functional decomposition as an early and integral part of systems engineering. My conclusion was that the G, T & E effort gave me the first and strongest idea as to how the big companies approached large-scale systems engineering tasks, with patience, perseverance, and good engineering judgment and with due regard for how to architect a large-scale system.

Meanwhile, I had maintained my contacts at the FAA and the DOT which led to some management and technical support contracts with the OST/DOT on the subjects of both the Civil Aviation Research and Development (CARD) study and the Climatic Impact Assessment Project (CIAP). And, as the NHTSA part of the DOT was expanding, I managed to obtain several contracts with them, largely in operations evaluation and research. So at that point I was working and studying, largely, in both systems engineering and operations research.

My technical contacts at the DOT led directly to about two years of study with the Technical Director of the Aviation Advisory Commission. In this capacity I provided immediate support on this wide-ranging investigation of the future of aviation in this country. It was truly a systems engineering look at current and future aviation systems, requiring a broad range of knowledge as well as a sense of synthesis, an important aspect of systems engineering. DOT contacts were quite useful in another direction. At about the time when the Dept of Energy was getting up and running, several professional contacts moved from the DOT to that new Department. Taking advantage of the people equation, I bid on and won an early contract with that Dept. It reinforced the fact that systems engineering was about the people as well as the subject.

Meanwhile, back in the early sixties, I remained an adjunct at GW and was finishing up my doctoral program. It culminated in a research activity with radar systems and the support of a key advisor, Solomon Kullback, as well as my primary advisor and Dept head, Nelson Grisamore. And as I took more responsibility at GW, I was asked to teach a certificate course in systems engineering. This reinforced my knowledge of systems engineering, and also led me to propose a master's program in systems engineering. The unique part of that program was its offering to companies on a cohort basis, rather than to students, one at a time. That turned out to be a very successful program for several industry participants, including Lockheed Martin and SAIC. The government agency known as the DMA also participated.

In another turn of events, and during a lull in my schedule, I took a trip to the Naval Systems Engineering facility at Port Hueneme. I made contact there with Capt. Wayne Meyer, and he responded well and immediately to my queries about how we might help him. That led to a contact to assist with his staff in terms of work on BPDSMS. This was a quite large Modeling and Simulation activity that was to help his staff maintain and improve that system. Here again, systems engineering came to the fore, along with a more serious introduction to the domain of shipboard air defense systems. That was to be very useful when it came time to think about how to enter the SDI program under the new Reagan missile defense strategy. Also, as part of that connection, the captain asked me to co-author a classified paper with him on the work with the BPDSMS. As it turned out, we had a strong capability with modeling that scenario which we quickly brought to the attention of RCA, who gave us a sole source contact to assist them in the SDI program. My task was overall management and special calculations of the lethal flux necessary to kill an enemy missile during its launch phase. That activity was a great challenge in terms of concept as well as all of its systems engineering thrusts.

As part and parcel of my job as an executive at Operations Research Inc (whose name had become simply ORI), it was necessary to "build the business". I did this by visiting with potential customers during lulls in my otherwise technical schedule. On one particular trip, the visit was close at hand at the aeronautics part of NASA Headquarters. As a result of that meeting I was given a "starter" sole source contract to attempt to assist my contact with aeronautics topics, which were largely supportive of what the country needed to do by way of aeronautics research that was not happening on its own in the industry. I would characterize that work as quick reaction support that required a broad range of interest and capability in aviation systems in general. As one of the deputy administrators of NASA in aeronautics declared during one assignment with him, "the objective is to cover my foxhole with paper". I thought that was one very good way of describing technical support activities, and still using the expression, from time to time.

Somewhere around 1986, I wrote a paper that was published in the C3I community Signal Magazine. My interest in systems engineering led me to conceive of and structure a book that I called "Computer-Aided Systems

Engineering". I was able to interest Prentice-Hall in that project and soon I had written, and they published, my first heavy-duty book (over 500 pages) on systems engineering. The Table of Contents of that book is as follows:

Chapter One – Systems Engineering and the Computer
Chapter Two – Formal Systems Engineering Structures
Chapter Three – Basic Computer Tools
Chapter Four – Diagramming Techniques
Chapter Five – Overview of Probability Concepts
Chapter Six – Specialized Computer Tools
Chapter Seven – Mathematical and Engineering Computer Tools
Chapter Eight – Information and Search Theory
Chapter Nine – Programming Languages
Chapter Ten – Top-Level Systems Engineering
Chapter Eleven – Scheduling and Costing
Chapter Twelve – Performance Modeling and Simulation
Chapter Thirteen – Risk and Decision Analysis
Chapter Fourteen – Software Development and Analysis
Chapter Fifteen – Systems Engineering Support Functions
Chapter Sixteen – Forecasting and Artificial Intelligence
Chapter Seventeen – Toward a Future Case Environment

This book stood me in good stead when the Dean of Engineering at GW offered me a full-time job as a full professor. It was then that I had to choose between staying in the industry or moving into academia. I soon chose the latter, the beginning of my 24-year stay at the GW Department of Engineering Management and Systems Engineering.

The job at GW required a primary course in systems engineering for which I took responsibility. As it turned out, after several iterations, the course outline evolved to the following:

WEEK	SUBJECT AREA
1	Overview of Systems Engineering
2	The 20 Elements of Systems Engineering
3	Requirements Engineering
4	Systems Architecting
5	Diagramming and Computer Tools
6	Applications of Probability Theory to Systems Engineering
7	Quantitative Relationships
8	Mid-Term Exam

(Continued)

WEEK	SUBJECT AREA
9	Scheduling and Costing
10	Performance Analysis/Simulation
11	Risk and Decision Analysis
12	Software Development I
13	Software Development II
14	Trends and Support Elements
15	Final Exam

BOOK WRITING IN SYSTEMS ENGINEERING

As I reached the age of 80, I decided that I was less mobile and therefore less able to carry out a variety of person-to-person systems engineering consulting tasks. I tried to fall back on writing as a third and more productive way to keep involved in the field. The result has been a relationship, over about five years, with the CRC Press and the writing and publishing of the following texts:

1. Thinking, An Approach to Systems Engineering Problem Solving
2. Systems Architecting, Methods, and Examples
3. Systems Engineering – 50 Lessons Learned
4. What Makes the Systems Engineer Successful? Surveys Reveal the Results
5. Problem-Solving, Leaning on New Thinking Skills
6. Cost-Effectiveness, A Systems Engineering Perspective
7. Tomorrow's Systems Engineering, Commentaries on the Profession

I have also been asked by INCOSE to present lectures, or colloquia, on several of these subjects. My overall strategy for book writing has been to keep busy each day, and not push myself beyond writing one to two pages each day. After a while I've got a hundred pages, which is my target for each book.

So I feel that I've been quite fortunate (lucky?) to have gotten to age 86 and I'm still kicking. Simple formula – a little bit of "work" each and every day, and still pay attention to the little things, and get them done. And pay attention to the family – they want the contact and still appreciate the connection. And you never know where the next piece of good work will come from. In my case, I've been doing physical therapy with Glenn for about a year, and we have decided to try a book together. Subject? A coach's book on basketball. We've got about 100 pages laid out, but so far have not pushed to get a publisher. But that will come in time. One step at a time, and a lot of patience. I'm following Dr. Duckworth's suggestion – persevere, each and every day. And if you miss a day or two, not to worry. You don't need the pressure of worrying.

So the tally goes as follows: three careers – 30 years as a working engineer, 24 years as a professor, and 9 years as a published writer. Not bad for an octogenarian. I've had a good run of it. Absolutely no regrets. And besides my technical career, I am able to add relationships with several women, the most spectacular of which was with June Linowitz, a person of great character and artist of some note and daughter of Sol Linowitz, previous ambassador to the OAS. On an occasional lunch with Sol, he would ask me to explain what operations research was. I tried, but never quite made it, to both of our satisfactions.

Transformations 12

This chapter deals with transformations that have occurred or are likely to occur, in both systems engineering and operations research. These transformations may or may not have previously been recognized, but according to this author, they have occurred, both in content as well as the method of delivery. We consider first the transformations in systems engineering

TRANSFORMATION NUMBER ONE – METHOD OF DELIVERY (ONLINE)

Delivery of systems engineering programs has evolved from face-to-face in-person delivery to numerous online representations. These online deliveries have been characterized by:

a. Fifteen sessions (for a typical course) in which each student is given an assignment whose response is evaluated, week by week, by the instructor

b. Interaction with each student is carried out on the internet rather than any face-to-face sessions

c. There may, or may not, be a capstone assignment representing the "final" exam

d. In general, team problem-solving is not used for online deliveries

e. It's a lot of work, especially for the instructor. It is estimated that the instructor needs to put in, each week, one to one and a half hours of interaction for each student. If he or she has a class of 30 students, that computes to 30 to 45 hours of work each week. That's a lot of work and it encourages "cutting corners" in various ways on the part of the instructor

DOI: 10.1201/9781003306610-12

Notwithstanding the above, online deliveries appear to be increasing in general and this author would not wish to change that trend. It seems to satisfy a demand and also respond to the hiring of adjunct professors for these courses.

TRANSFORMATION NUMBER TWO – METHOD OF DELIVERY (COHORT-BASED)

Another distinct method of delivery is moving into cohort-based delivery. Under this method, the following appears to be true:

A cohort of some number (often 30–40) students in one company are accepted for a degree (often a master's) and courses are provided to the site of the location of this cohort, with face-to-face teaching. This is especially convenient for both the student as well as the instructor and has been a very successful method of delivery.

TRANSFORMATION NUMBER THREE – CONTENT

This change has evolved largely inadvertently through the course of committee work within INCOSE and the standards makers. The driving force appears to have been standard 15288 [1] which defined systems engineering in terms of the following processes:

a. Technical processes
b. Technical management processes
c. Agreement processes
d. Organizational project-enabling processes

Looking at these four types of processes, we see that three of them are management-oriented. To this author, that represents a significant change away from purely technical to management. This was pointed out by an INCOSE fellow [2], in a paper that asked the question "where is the engineering?" As far as this author is concerned, this question has not been addressed by INCOSE which would be a natural way to deal with it.

This author largely agrees that the transformation has yet to occur and offers the following seven areas in which the technical aspects of systems engineering could be added:

a. Tradeoff analysis
b. Definition of alternatives in design
c. Interoperability analysis
d. New and specific methods of system architecting
e. Cost-effectiveness analysis
f. Requirements satisfaction analysis
g. Single-point failure analysis

Tradeoff analysis: This has been accepted as a critical part of systems engineering. The process involves looking at two or more design alternatives and trading one against the other to select the best alternative.

Definition of alternatives in design: This is the process of identifying what constitutes a viable alternative.

Interoperability analysis: Recall that when constructing the synthesis chart, we list the various design approaches for each function. This process involves looking down the rows and checking to assure that there are no interoperability problems, i.e., each alternative will be able to interoperate with the other alternatives.

Cost-effectiveness analysis: We are told by various customers (the Military, for example) that we are in search of a cost-effective solution. For this activity, we are doing exactly that.

Requirements satisfaction: As we construct alternative architectures, we need to check to make sure that we indeed satisfy all the requirements that we have agreed upon with the customer.

TRANSFORMATION NUMBER FOUR – PROBLEM SOLVING

This transformation has been suggested, but not fully accepted as necessary. It evolves from a panel discussion some years ago [3,4] by a distinguished group of systems engineers that addressed the issue from the perspective of "learning from failure". The main points made by this panel were:

a. There were several cases in which NASA literally "learned from failure" which this author considered to be transformational. This learning occurred for the following systems: the X-33 Reusable Launch Vehicle, the Hubble Space Telescope, and the Delta 180 Powered Space Intercept

b. The panel also concluded that "more process was not the solution". This is transformational in the sense that accepting that conclusion would lead to rejecting that solution institutionally. That is an important conclusion that, if accepted, leads to change in behavior of some import.

TRANSFORMATION NUMBER FIVE – MANAGEMENT

When Dan Goldin became the administrator of NASA in 1992, he declared that a serious performance goal was to build systems faster, cheaper, and better. That was revolutionary (transformative?) if it could be achieved. Many thought that it was possible to achieve two of three, but not all three. With improved management, said Goldin, all three could be achieved. This author assumes that he was able to do so (achieve all three) on selected systems developments. For that reason, we stretch a point and include this as one of our transformations.

Goldin, according to many, was transformational in his approach to NASA's charter. His well-known approach was to visualize the result that he wanted and then work backward to achieve that result. He quoted Michaelangelo with the saying "I see the angel in the marble and then I carve until I set it free. This has become to be a Zen-like approach in more modern days. This author likes to think that Goldin's approach was transformational during his time as an administrator in that he influenced his management team and those who were able to appreciate what he was saying as the "boss".

TRANSFORMATION NUMBER SIX – "ELEGANT" PRODUCTS

This transformation has also been suggested, but not formally accepted as necessary. It was suggested in a paper by Michael Griffin, previously an administrator of NASA. That change in his address as to how to "fix" systems engineering [5]. Griffin declared "in academia and advanced research, I believe we can first ask interesting questions. From there we can set up experiments and studies to find the answers" to get to systems that are "elegant". Griffin, with his insight and approach, concluded that it was not a matter of insufficient process, which tended to change the way in which many thought progress could be achieved.

TRANSFORMATION NUMBER SEVEN – ARCHITECTING SYSTEMS USING COST-EFFECTIVENESS METHODS

The approach to architecting systems has been largely dominated by what has been called the DoDAF (Department of Defense Architectural Framework). Systems engineering has undergone significant growth over the years. That growth is exemplified by the difference between conceptions in the technical features of systems engineering as seen by A.D. Hall back in 1965, and the way it has evolved as represented in the INCOSE handbook, number four. A summary of the two conceptions is shown in the Table below – to show the similarities and differences explicitly (Table 12.1).

TABLE 12.1 Similarities and Differences in the Technical Aspects of Systems Engineering

TOPICS IN A.D. HALL [6]	TOPICS IN INCOSE HANDBOOK (LISTED AS "TECHNICAL PROCESSES") [7]
The Concept of Planning	Mission Analysis
The Five Phases of Systems Engineering	Stakeholder Needs and Requirements definition
Problem-Solving Models	System Requirements Process
An approach to Problem Definition	Architecture Definition process
Input-Output Technique	Design Definition Process
Systems Synthesis	System Analysis Process
Systems Analysis	Implementation Process
Selecting the Optimum Systems	Integration Process
Systems Development	Verification Process
Decision Making	Transition Process
Economic Theory of Value	Validation Process
Creative Thinking	Operation Process
Graphical Models	Maintenance Process
	Disposal Process

Yesterday's systems engineering can be described by Hall as well as Goode and Machol back some 30 years ago.

The Department of Defense Architectural Framework (DoDAF): There has been considerable unhappiness with the various aspects of the DoDAF approach, but an alternative has not been adopted by the DoD. Back in 2020, this author set forth [8–10] a quite specific method of system architecting that involves the following four concrete steps:

1. Functional decomposition
2. Synthesis
3. Analysis
4. Selection of most cost-effective architecture

TRANSFORMATION NUMBER EIGHT – MOVEMENT TOWARD REUSABLE SOFTWARE SYSTEMS

Many have noted the fact that the software engineering subset of systems engineering remains an art, with many difficulties in bringing these systems in within cost, schedule, and performance goals. A proposed solution has been to start a new initiative in the domain of reusable whole software systems (DOTSS). This approach promises a transformation in how to acquire "new" software systems and has been proven by the examples of Microsoft Office and Lotus's SmartSuite.

TRANSFORMATION NUMBER NINE – ACCEPTANCE OF MODULAR SYSTEMS APPROACH (MOSA)

This is a new way of approaching the design of large-scale systems, considered an opportunity for collaboration between DoD and Industry. It holds out promises for more effective systems in terms of cost, schedule and performance.

We now move on to the field of operations research and ask ourselves as to the nature of transformations that have occurred there.

TRANSFORMATION NUMBER TEN – ONLINE CLASSES IN OPERATIONS RESEARCH

This is the same as transformation number one, except that it has occurred in operations research vs. in systems engineering.

TRANSFORMATION NUMBER ELEVEN – CONTENT IN OPERATIONS RESEARCH

We can infer transformation by placing content features from Morse and Kimball with features of more up-to-date representations (as per Hillier and Lieberman). This is shown in Table 12.2 below:

TABLE 12.2 Comparison of Old and New Content in Operations Research

OLD CONTENT (MORSE AND KIMBALL)	NEW CONTENT [HILLIER AND LIEBERMAN [7]]
Probability	Mathematical Programming
Measures of Effectiveness	Network Analysis
Strategical Kinematics	Game Theory
Tactical Analysis	Probability Theory
Gunnery and Bombardment	Queuing Theory
Operational Experiments	Inventory Theory
Organizational and Procedural Problems	Reliability
	Decision Analysis
	Simulation
	Advanced Topics

Looking at the above table, we note the change from analysis in the military domains to a more specific theory, not necessarily applied to any one field. Examples are reliability and inventory theory. So, as they say, operations research has come a long way from the early days to today. Mathematical constructs, of various types, have found their way increasingly into the content of operations research.

TRANSFORMATION NUMBER TWELVE – ANALYTICS

Some operations search administrators are including the topics of "analytics" into the field. In particular, this author has noticed that the London School of Economics has added analytics to their program in operations research. This, perhaps, is because the overall field is fitting into business schools and these schools now embrace analytics as a serious business subject. Whatever the reason, analytics has come of age, and a convenient location for it is to find a spot in operations research, which, in turn, resides in a business school.

What is analytics? [11] According to several schools, it is a form of data modeling for which there appear to be four types: descriptive, diagnostics, predictive and prescriptive. A very recent issue of ORSA today [12] defines business analytics as a multidisciplinary activity that lies at the intersection of computer science, statistics, mathematics (esp. optimization), and business.

A very recent issue of INFORMS Today [13] reports on HP's use of analytics whereby its Data Science and Knowledge Discovery (DSKD) affinity group, composed of more than 3000 members held a summit with 114 presentations over three days with quite varied content. HP is devoted to using technology in the service of humanity and using analytics across various groups with great success. They are now (2021) at a revenue level of 463.5B and a leader with 37,000 patents. With its strength and perseverance, HP can "lead the charge" into analytics infusions and research such as to be recognizable as truly transformative in the field of operations research.

TRANSFORMATION NUMBER THIRTEEN – WIDE AND DEEP PROGRAMMING

Another observation regarding the content of today's operations research is to acknowledge that the mathematical programming topic actually contains several sub-topics, including linear programming, goal programming, dynamic programming, integer programming, and non-linear programming [6]. Thus, programming is a rather broad term that is both wide and deep.

TRANSFORMATION NUMBER FOURTEEN – OPTIMIZATION THEORY

Within the overall topic, the research analyst often wishes to develop an optimum. For this purpose, he or she will go to software that will allow for optimization by means of its internal algorithm. Several packages claim to be representative of optimization. They include the following list [14]:

- Brandfolder
- Bynder
- Media Valet
- Frontify
- Canto
- merlinOne
- Imagen
- NetX
- Cortex

TRANSFORMATION NUMBER FIFTEEN – ROAD-MAPPING

A relatively recent [15] issue of the IEEE Transaction on Engineering Management focused on the topic of road-mapping, which reminded this author of its significance. Road-mapping pointed to the fact that road-mapping emerged from industrial practice "some five decades ago" and is alive and well today. Road-mapping is a visual forward-looking document that communicates both tactical and strategic plans in a particular business area (sector). This article declares that the first sector-level application of road-mapping was in 1991, in the arena of semiconductors (the International Technology Roadmap for Semiconductors [ITRS]). Roadmaps contain a compact method of visualizing plans of the future, emphasizing temporal representations.

TRANSFORMATION NUMBER SIXTEEN – PROGRAM EVALUATION AND REVIEW TECHNIQUE [16]

PERT (the Program Evaluation and Review Technique) is a network-based schedule that explicitly shows all dependencies between tasks in an overall schedule. It was developed under the auspices of the Polaris program and was considered new and definitive as far as scheduling is concerned. Prior to PERT, Gantt charts were the preferred method of scheduling. A list of PERT software for project planning is provided as follows:

1. Monday.com (*)
2. Wrike
3. Workzone
4. Asama (*)
5. Zoho Projects
6. Kintone
7. Smartsheet (*)
8. Function Fox
9. Swift Enterprise
10. Jira
11. ProWorkFlow
12. Caspio
13. Knack
14. Airtable 980
15. ClickUp (*)

(*) designated as superior in overall performance

TRANSFORMATION NUMBER SEVENTEEN – DIGITAL ENGINEERING [17]

- This transformation is occurring in the battlespace, (speed, cycle time)
- Need to reform the department for greater performance and affordability

- Requires a new vision for the way we conceive, build, test, and sustain our national defense systems
- Reforms business practices by connecting people, processes, data, and capabilities
- It combines model-based and digital practices
- Complete reformation of practices across the board in the DoD

REFERENCES

1. Standard ISO/IEC/15288
2. Wasson, C. "The Systems Engineering Conundrum: Where is the Engineering? See https://online library.wiley.com
3. Slegers, N., et. al. "Learning From Failure in Systems Engineering: A Panel Discussion", see Systems Engineering, DOI
4. See https://seaver.pepperdine.edu/newsroom
5. Griffin. M., "Michael Griffin explains How to fix systems engineering", Stevens Institute of Technology, talk, 14 December 2010
6. Hall, A.D., "A Methodology of Systems Engineering", D. Van Nostrand, 1962
7. Walden, D., et. al., "Systems Engineering Handbook", fourth edition, John Wiley, 2015
8. Eisner, H., "Systems Architecting", CRC Press, 2019
9. Capterra, see en.wikipedia/capterra, DAM Software
10. Hillier, F., and G. Lieberman, "Introduction to Operations Research:", Third Edition, Holden-Day, 1980
11. See www.analytics8.com
12. See https://en.wikipedia.org/wiki/Mathematical_optimizationAdvanced
13. Curtland, C., P. Neto, and A. Ghozeil, "HP's Advanced Analytics Powders Technology in the Service of Humanity", INFORMS Today, Volume 49, Number 2, April 2022
14. Phaal, P., and C . Kerr, "Guest Editorial: New Perspectives on Road Mapping: Foreword "IEEE Transactions on Engineering Management:, Volume 69, Number 1, February 2022
15. See Hillier and Lieberman, chapter 6, Section 6.6
16. PERT, Program Evaluation and Review Technique, Operations Research, 1959
17. Zimmerman, P., Office of the Under Secretary of defense for research and engineering, a model-Based enterprise Summit Digital Enterprise Transformation; SERC Research Program, SERC, Hoboken New Jersey; Digital Engineering Strategy and Implementation, NIST, Model-Based Enterprise Summit 2019, April 2019

Summary

13

This chapter provides a brief summary of the previous twelve chapters.

Chapter 1 is a very brief, single-page, introduction.

Chapter 2 provides an overview of the field of operations research. This includes the three seminal texts from Morse and Kimball, Hillier and Lieberman, and Churchman et al. Subject matter is identified as solution domains and tools. Selected Departments of Operation Research are cited along with four dominant courses from the GWU. The transition to military applications is noted to include MORS. Well-known personnel are also cited, together with the British influence on the field.

This next chapter (3) contains a brief overview of the field of systems engineering, starting with Standard 15288 and the processes defined therein. This is followed by an articulation of the "30 elements of systems engineering", as defined in an earlier text from this author. Then we look at the elements of systems engineering as represented by A.D. Hall, Goode and Machol, and Machol in his handbook. A historical perspective is offered as defined in the INCOSE handbook. An interesting footnote is provided by Michael Griffin, and a set of heuristics as suggested by Rechtin in his treatise on system architecting. From there, information from the DoD is provided, along with eight core areas for better buying power. Opportunities for collaboration between the DoD and Industry are articulated, along with a short glimpse into the future (from INCOSE) as well as a suggestion for a grand unified theory (GUT) of systems engineering.

Chapter 4 provides representations from academia, for both systems engineering as well as operation research.

In chapter 5, we find a visitation into system engineering, with a reiteration of the systems approach as well as Griffin's suggestion as to how to fix systems engineering. Next, a senior group of systems engineers takes on the subject of "how to fail in systems engineering, and lessons thereof. Fellow Charles Wasson raises the question "where is the engineering" in systems engineering, followed by an overview of the MITRE approach to systems engineering as well as that of NASA.

DOI: 10.1201/9781003306610-13

Chapter 6 articulates some special topics in operations research with emphasis on military applications, moving into the topics of soft OR and Checkland's soft systems methodology. Then we define the acronym CATWOE and reiterate the goals for 2022 from the new president, Radhika Kulkarni.

The next chapter (7) defines some common elements of both systems engineering and operations research, namely:

   a. Modeling and simulation
   b. Optimization theories
   c. Software for the above
   d. Probability and statistics applications
   e. Reliability theory
   f. Decision theory
   g. Cost-effectiveness analysis
   h. Forecasting
   i. Management

Chapter 4 contains various representations of programs in academia in both operations research as well as systems engineering.

The next chapter (8) tackles the elusive topic of growth in both systems engineering as well as operations research. Growth is defined as the evolution of capability in terms of being able to deal with the following:

   a. Local (project) management
   b. Local leadership
   c. Problem-solving
   d. Systems architecting
   e. Enterprise architecting

In chapter 9, we find a list of key contributions in operations research. This is done by topic as well as by leaders (as many as 30) in each field.

Chapter 10 cites several key contributions to systems engineering, in a manner analogous to the chapter on operations research. A list of some 20 key contributors is shown.

In chapter 11, we get to lean in on some personal history of the author in relation to both operations research as well as systems engineering.

Chapter 12 articulates some seventeen transformations that are worth noting, in both operations research as well as systems engineering.

Index

Printed in the United States
by Baker & Taylor Publisher Services

Printed in the United States
by Baker & Taylor Publisher Services